Student Companion with Complete Solutions for
An Introduction to Genetic Analysis

FOURTH EDITION

Student Companion with Complete Solutions for

An Introduction to Genetic Analysis

FOURTH EDITION

by
David T. Suzuki, Anthony J. F. Griffiths,
Jeffrey Miller, and Richard C. Lewontin

Diane K. Lavett

State University of New York at Cortland

W. H. Freeman and Company
New York

Printed in the United States of America

3 4 5 6 7 8 9 0 VB 9 9 8 7 6 5 4 3 2 1 0

Table of Contents

	Preface	vii
Introduction:	**How to Think Like a Geneticist**	viii
	How to Solve Problems in Genetics	ix
	How to Study Genetics	ix
Chapter 1:	**Genetics and the Organism**	1
Chapter 2:	**Mendelian Analysis**	4
	Working With Probability	5
Chapter 3:	**Chromosome Theory of Inheritance**	22
Chapter 4:	**Extensions to Mendelian Analysis**	34
	A Systematic Approach to Problem Solving	69
Chapter 5:	**Linkage I: Basic Eukaryotic Chromosome Mapping**	73
Chapter 6:	**Linkage II: Special Eukaryotic Chromosome Mapping Techniques**	98
	Some Tough Advice on How to Study Genetics	127
Chapter 7:	**Gene Mutation**	128
Chapter 8:	**Chromosome Mutation I: Changes in Chromosome Structure**	136
Chapter 9:	**Chromosome Mutation II: Changes in Chromosome Number**	154
Chapter 10:	**Recombination in Bacteria and Their Viruses**	164
Chapter 11:	**The Structure of DNA**	177
Chapter 12:	**The Nature of the Gene**	185
Chapter 13:	**DNA Function**	199
Chapter 14:	**The Structure and Function of Chromosomes in Eukaryotes**	207
Chapter 15:	**Manipulation of DNA**	213
Chapter 16:	**Control of Gene Expression**	224
Chapter 17:	**Mechanisms of Genetic Change I: Gene Mutation**	230
Chapter 18:	**Mechanisms of Genetic Change II: Recombination**	237
Chapter 19:	**Mechanisms of Genetic Change III: Transposable Genetic Elements**	245
Chapter 20:	**The Extranuclear Genome**	250
Chapter 21:	**Genes and Differentiation**	261
Chapter 22:	**Genetic Analysis of Development: Case Studies**	269
Chapter 23:	**Quantitative Genetics**	275
Chapter 24:	**Population Genetics**	287

Preface

Unlike many solutions manuals, *Student Companion with Complete Solutions for An Introduction to Genetic Analysis, Fourth Edition* attempts to provide a logical approach to solving genetics problems. While explaining the reasoning behind each answer given, the book also recognizes that what is obvious to geneticists is seldom apparent to the beginning student.

The *Companion* was written from the perspective of a teacher who is standing in front of a blackboard, trying to explain a problem to a class of beginning genetics students. Because I have been in that position many times, my students have taught me exactly where it is that difficulties will occur. I have tried to anticipate each of those potential obstacles in the explanations that follow.

The *Companion* has been tested in classrooms at the State University of New York at Cortland and Emory University. In addition, Dr. Anthony J. Pelletier of the Department of Molecular, Cellular, and Developmental Biology of the University of Colorado at Boulder has independently worked all the problems presented in the text. His careful, thorough work resulted in the detection of many errors, and he has my deepest respect and gratitude for his magnificent effort. Although Dr. Pelletier provided this invaluable service, he should not be held responsible for any errors that may still remain. Those errors are mine, and I hope that all users of the *Companion* will feel free to communicate directly with me about any mistakes that they detect.

Diane K. Lavett
Department of Biological Sciences
The State University of New York at Cortland

Introduction: How to Think Like A Geneticist

You will sharpen a number of skills as you work through your textbook and the *Companion*. Since genetics requires very careful reading, you should become a more careful reader by the end of your course of study. For example, a great deal of information is conveyed by the sentence, "Two mutants were crossed and a wild-type phenotype was observed in the male offspring." If you doubt the amount of data meant to be conveyed by that sentence, I will list it, even though the information may have little meaning for you until later:

1. Two separate genes are involved.
2. Both mutants are recessive.
3. The mutant gene in the female is located on an autosome.

There are seldom superfluous words in a genetics problem. Consequently, you must learn to think about each word that is provided.

A second skill that will be sharpened as you progress through this course is systematic thought. To approach a problem in genetics in a haphazard manner is to enter quickly into the realm of chaos and confusion. I hope that this book will be of real assistance in sharpening your ability to think systematically.

A third skill that you should learn is what I call being gentle with your so-called mistakes. Our educational system has labelled as a mistake the failure to arrive at the correct answer on the first try. Yet in genetics, as in all of science, progress is made by learning through trial and error what the explanation is *not*. When you think in this way, I hope that you will begin to view your attempts to solve a problem as hypotheses that are being rejected rather than as errors. If you are not able to answer a particular problem, you need only to go back to your initial assumptions and revise them. Perhaps you have misread the problem. Perhaps you have correctly thought but have made a simple mathematical error. An important point to keep in mind is that in any trial and error learning, there often must be a number of "trials." The only true mistake that you can make is to stop generating hypotheses and to give up.

How to Solve Problems in Genetics

Genetics is not a spectator sport; you cannot learn genetics without solving problems. A serious threat to your ability to learn genetics is misuse of the *Companion*. You can convince yourself far too easily that you understand a problem as you read the solution to it when, in fact, you do not understand it at all. Let me suggest the proper way to use this book:

1. Read the section entitled "Important Terms and Concepts" before beginning to work the problems at the end of a chapter. If any term or concept does not cause you to recall exactly what the text said about it, reread that section of the text.

2. Work on a problem without reading the *Companion* until you are truly stuck.

3. Read the explanation of the solution.

4. Without consulting the *Companion*, immediately rework the problem.

5. Two or three days later, work the problem without consulting the *Companion*. If you cannot do it at this point, you probably did not understand the problem earlier.

6. If you cannot work the problem without consulting the *Companion*, repeat steps 2 to 4 once. If you cannot work the problem at that point, consult your teacher or a friend who can explain the problem to you.

7. Throughout the problem-solving process, consult the section entitled "Tips on Problem Solving" as needed. There is a limited number of types of problems. For each type of problem, there is a pattern to the method of solution. Each problem solved in this book has followed the pattern best suited for the problem. Learn to duplicate the patterns.

Once you have mastered all problems from the text, take the Self-Test found at the end of each chapter. If you correctly solve all of these problems without referring to the answers supplied, you should be in good position to handle whatever your teacher may ask you in a test.

How to Study Genetics

The process of learning outlined above cannot be completed in a "cram" session the night before a test. The attempt to learn genetics in that fashion is doomed to failure. The best approach is to study genetics each day, weekends included. Keep your study sessions short, not more than two or three hours at a time. Study in a quiet place where you will not be interrupted or distracted. If you find your concentration wavering, take a short break. If you find that anxiety is interfering, go for a run. Avoid caffeine, both while studying and, most importantly, before taking a test. If, on the night before a test, you are forced to make a choice between a good night's sleep and trying to learn far too much material for the time available, choose sleep.

Many students will have some difficulty with the material and problems in Chapters 2 through 6. Thereafter, the material will be easier to conceptualize, and the problems will be easier to do. The reason for the difficulty with the earlier chapters is that they require a level of abstract thought not usually demanded in undergraduate courses. Beginning with Chapter 7, however, the material becomes more descriptive and, simultaneously, more consistent with the skills required for success in other biology courses. Be aware that you will have to work quite hard in dealing with this early material; also be aware that you are not alone in your difficulty.

Generations of students have struggled with Mendelian genetics, and the vast majority have been successful in their effort. Their reward has been that they have learned a new way to view the universe. It is my sincere hope that this will be your reward, too.

CHAPTER

1

Genetics and the Organism

Important Terms and Concepts

Genetics is the study of the inheritance of traits by means of the examination of their variation.

Genes are the basic functional units of heredity. They are composed of DNA and they determine specific traits.

Chromosomes consist of long DNA molecules complexed with protein. They contain many genes.

Transmission genetics is the study of the ways in which genes are passed from generation to generation.

Molecular genetics is the study of gene structure and function.

Population genetics is the study of gene behavior in populations.

The life of any particular organism results from the interaction of its inherited material with its environment.

The **genotype** refers to the inherited genes in an organism.

The **phenotype** refers to the physical appearance of an organism.

The **norm of reaction** refers to the environment-phenotype relationship for a specific genotype.

Developmental noise is the random variation that occurs in phenotype when both genotype and environment are held constant.

Self-Test

1. What does the abbreviation "DNA" mean?

2. How many separate DNA molecules are there in an organism that has ten pairs of chromosomes, each with one chromatid per chromosome?

3. What are the three major subdivisions of genetics? Define each.

4. Is the DNA in your father identical to the DNA in your mother?

5. Is your DNA totally different from the DNA found in a dog? An oak tree?

6. Which is more important to an organism, its genetic information or the environment in which it develops?

7. If two genetically identical organisms develop in identical environments, will they necessarily be identical for a specific trait?

8. If two genetically different organisms develop in identical environments, will they necessarily be nonidentical for a specific trait?

9. If two genetically different organisms develop in different environments, will they necessarily be nonidentical for a specific trait?

10. Distinguish between genotype and phenotype.

Solutions to Self-Test

1. Deoxyribose nucleic acid.

2. Because each chromosome contains one long DNA molecule, there would be $10 \times 2 = 20$ DNA molecules in the organism.

3. Transmission genetics is the study of gene transmission from generation to generation. Molecular genetics is the study of gene structure and function. Population genetics is the study of gene behavior in populations.

4. No. Although your father and mother have genes that control the same functions, not all the genes are identical. Because the genes are composed of DNA, not all the DNA would be identical.

5. No. All organisms must solve similar problems in order to survive. As an example, the Krebs cycle occurs in humans, dogs, and trees. Because these metabolic changes take place in all three organisms, identical or very similar enzymes must exist in all three organisms. Because the information for the enzymes resides in the DNA, the DNA responsible for these enzymes must be identical or very similar.

6. Both are very important. The inherited material initially sets the limits for the organism, while the development of an organism in a specific environment determines where along the spectrum of the possible the organism will be.

7. No.

8. No.

9. No.

10. Genotype is the inherited information, while phenotype is the end result of an interaction of that information with the environment.

CHAPTER

2

Mendelian Analysis

Important Terms and Concepts

A **gene** controls one characteristic, or trait. **Alleles** are alternative forms of a gene. A good analogy is a coin. You can have pennies, nickels, and dimes, all "genes." You can have 1956, 1971, and 1989 pennies, all "alleles" of the penny "gene." Alleles determine alternate types (long, short) for the gene (height) being studied.

A **pure line** is a strain that breeds true for the trait being studied. It is also called a **true breeding line**.

The **parental generation**, P, is the generation from which the first cross in a series is taken. Usually, but not always, the parents are true breeding.

The **first filial generation**, F_1, consists of the progeny from the parental generation.

Dominant refers to an allele that is expressed regardless of the other alleles present for the trait being studied. **Recessive** refers to an allele that is expressed only when it is the only type of allele present in an organism for the trait being studied.

Equal segregation of gene pairs refers to the separation of alleles into gametes, which contain only one allele each. This is the **first law of Mendel.**

A **heterozygous** line, or **hybrid** line, contains two different alleles for the trait being studied.

A **homozygous** line contains identical alleles for the trait being studied. The line may be either homozygous recessive or homozygous dominant, depending upon which allele is present.

Monohybrid refers to a single gene pair. F_1 monohybrid crosses lead to 1:2:1 ratios.

Dihybrid refers to two gene pairs being studied simultaneously. F_1 dihybrid crosses result in a 9:3:3:1 ratio.

Independent assortment of gene pairs refers to equal segregation of one allelic pair independently of another allelic pair. This is the **second law of Mendel.**

A **testcross** is a cross of a homozygous recessive organism with an organism of dominant appearance. It results in a 1:1 ratio if the organism is heterozygous and a 1:0 ratio if it is homozygous.

Pedigree analysis is a tool for determining the genetic status of related individuals over several generations.

The **propositus** is the individual that first called attention to the family being studied.

Working with Probability

Probability is the number of times an event is expected to happen divided by the total number of times it could have happened.

"And" statement: indicates the need to multiply.

Example: the probability of a black female horse

= p(black horse) "and" the p(female horse)

= p(black horse) × p(female horse)

Example: the probability of *AA Bb*

= p(*A* "and" *A*) "and" p(*B* "and" *b*)

= p(*A* × *A*) × p(*B* × *b*)

"Or" statement: indicates the need to add.

Example: the probability of a black or brown horse

= p(black horse) "or" p(brown horse)

= p(black horse) + p(brown horse)

Example: the probability of *AA* or *Aa*

= p(*A* "and" *A*) "or" p(*A* "and" *a*)

= $p(A \times A) + p(A \times a)$

"At least" statement: indicates the need to add.

Example: the probability of at least three out of five

= p(3 out of 5) "or" p(4 out of 5) "or" p(5 out of 5)

= p(3 out of 5) + p(4 out of 5) + p(5 out of 5)

Remember that the probability of at least three out of five = 1 - the probability of 0 or 1 or 2 out of 5.

Be sure that you have thoroughly read the entire chapter before you attempt any of the problems.

Solutions to Problems

1. Mendel's first law states that alleles segregate during meiosis. Mendel's second law states that genes independently assort during meiosis.

2. To determine whether the *Drosophila* is *AA* or *Aa*, a testcross should be done. By definition, this means using a fly that is *aa*. If the original fly is *AA*, all progeny will have the A phenotype. If the original fly is *Aa*, one half the progeny will have the A phenotype and one half will have the aa phenotype.

3. The progeny ratio is 3:1, indicating a classical heterozygous by heterozygous mating. The parents must be $Bb \times Bb$. Their black progeny must be *BB* and *Bb* in a 1:2 ratio, and their white progeny must be *bb*.

4. Begin by drawing a pedigree, letting the normal allele be *T* and the Tay-Sachs allele be *t*.

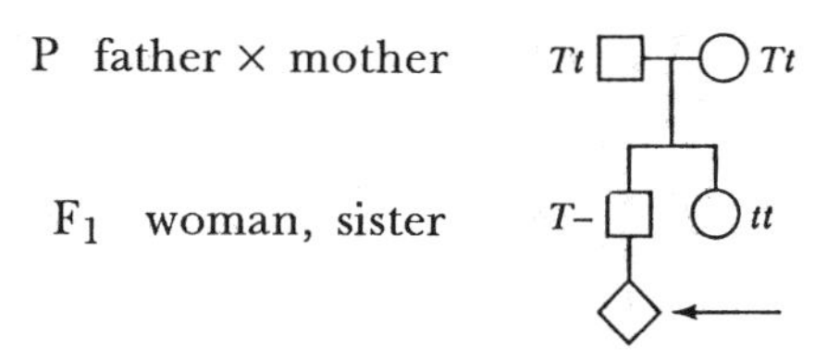

The woman's sister has to be *tt* because she had Tay-Sachs. Therefore, both parents are carriers, and the cross was $Tt \times Tt$. The woman of concern here is either *TT* (probability of 1/3) or *Tt* (probability of 2/3).

If she marries a man who is also a carrier, their unborn child has a probability of 1/4 of having Tay-Sachs, given that the woman is a carrier. Thus, the final probability of having a child with Tay-Sachs is

$$p(\text{woman is } Tt) \times p(tt \text{ child from } Tt \times Tt \text{ mating})$$
$$= (2/3)(1/4) = 2/12 = 1/6$$

If she marries a man who is not a carrier, then she will not have a child with Tay-Sachs.

If she is a carrier, each of her normal children has a 50% chance of being a carrier, no matter whom she marries.

The woman should be informed of all the probabilities, including the probability that her mate will be a carrier, which will vary drastically with ethnic group, and she should be encouraged to have the carrier status checked for herself, her mate, and all normal children. She should be told that her overall risk of having a Tay-Sachs child is slight.

5. Each die has six sides, so the probability of any one side (number) is 1/6. To get specific red, green, and blue numbers involves "and" statements.

a. $(1/6)(1/6)(1/6) = (1/6)^3$

b. $(1/6)(1/6)(1/6) = (1/6)^3$

c. $(1/6)(1/6)(1/6) = (1/6)^3$

d. To get no sixes is the same as getting anything but sixes:

$(1 - 1/6)(1 - 1/6)(1 - 1/6) = (5/6)^3$.

e. There are three ways to get two sixes and one five:

6R, 6G, 5B	(1/6)(1/6)(1/6)
or	+
6R, 5G, 6B	(1/6)(1/6)(1/6)
or	+
5R, 6G, 6B	(1/6)(1/6)(1/6)
	$3(1/6)^3$

f. Here there are "and" and "or" statements:

$$p(\text{three sixes or three fives}) = (1/6)^3 + (1/6)^3 = 2(1/6)^3$$

g. There are six ways to fulfill this:

$$6(1/6)^3 = (1/6)^2$$

h. The easiest way to approach this problem is to consider each die separately:

The first die thrown can be any number. Therefore, the probability for it is 1.0.

The second die can be any number except the number obtained on the first die. Therefore, the probability of not duplicating the first die is 1.0 - p(first die duplicated) = 1.0 - 1/6 = 5/6.

The third die can be any number except the numbers obtained on the first two dice. Therefore, the probability is 1.0 - p(first two dice duplicated) = 1.0 - 2/6 = 4/6.

Therefore, the probability of all different dice is (1.0)(5/6)(4/6) = 20/36 = 5/9.

6. a. Before beginning the specific problems, write the probabilities associated with each jar.

jar 1	p(R)	= 600/(600 + 400) = 0.6
	p(W)	= 400/(600 + 400) = 0.4
jar 2	p(B)	= 900/(900 + 100) = 0.9
	p(W)	= 100/(900 + 100) = 0.1
jar 3	p(G)	= 10/(10 + 990) = 0.01
	p(W)	= 990/(10 + 990) = 0.99

1. p(R, B, G) = (0.6)(0.9)(0.01) = 0.0054
2. p(W, W, W) = (0.4)(0.1)(0.99) = 0.0396

3. Before plugging into the formula, you should realize that, while white can come from any jar, red and green must come from specific jars (jar 1 and jar 3). Therefore, white must come from jar 2:

$$p(R, W, G) = (0.6)(0.1)(0.01) = 0.0006$$

4. $p(R, W, W) = (0.6)(0.1)(0.99) = 0.0594$

5. There are three ways to satisfy this:

$$R, W, W \quad or \quad W, B, W \quad or \quad W, W, G$$
$$= (0.6)(0.1)(0.99) + (0.4)(0.9)(0.99) + (0.4)(0.1)(0.01)$$
$$= 0.0594 + 0.3564 + 0.0004 = 0.4162$$

6. At least one white is the same as 1 minus no whites:

$$p(\text{at least 1 W}) = 1 - p(\text{no W}) = 1 - p(R, B, G)$$
$$= 1 - (0.6)(0.9)(0.01) = 1 - 0.0054 = 0.9946$$

b. The cross is $Rr \times Rr$. The probability of red (R-) is 3/4, and the probability of white (rr) is 1/4. Because only one white is needed, the only unacceptable result is all red.

In n trials, the probability of all red is $(3/4)^n$. Because the probability of failure must be 5 percent

$$(3/4)^n = 0.05$$
$$n = 10.41, \text{ or 11 times.}$$

7. Charlie, his mate, or both, obviously were not pure-breeding, because his F_2 progeny were of two phenotypes. Let black and white = A, red and white = a. If both parents were heterozygous, then red and white would have been expected in the F_1 generation. It was not observed, so only one of the parents was heterozygous. The cross is

$$P \quad Aa \times AA$$
$$F_1 \quad 1\,Aa : 1\,AA$$

Two F_1 heterozygotes (Aa) when crossed would give 1 AA (black and white) : 2 Aa (black and white) : 1 aa (red and white).

If the red and white F_2 progeny were from more than one mate of Charlie's, then the farmer acted correctly. However, if the F_2 progeny came only from one mate, the farmer may have acted too quickly.

8. You are told that normal parents have affected offspring. This is the pattern with recessive disorders.

9. Many different approaches could be used. What follows is one approach. An asterisk indicates that a decision as to dominance can be made if that result occurs.

Cross 1: white × white

* 1. if blue appears, white is dominant

2. if all white, one of the parents is homozygous

Cross 2 blue × blue

* 1. if white appears, blue is dominant

2. if all blue, one of the parents is homozygous

If each cross reveals a homozygous parent,

Cross 3: white × blue (one each from parents of Crosses 1 and 2)

* 1. if 100% one color, that color is dominant

2. if 1:1 ratio, one of the parents is heterozygous

If one of the parents is heterozygous in Cross 3,

Cross 4: white progeny from Cross 3 × white parent from Cross 3

1. 100% white, white is recessive or one parent is homozygous

* 2. if some blue, white is dominant

Cross 5: blue progeny from Cross 3 × blue parent from Cross 3

1. 100% blue, blue is recessive or one parent is homozygous

* 2. if some white, blue is dominant

10. Because some children of identical parents have a different phenotype, the parents are heterozygous and the trait is dominant.

11. To do this problem, you should first draw a pedigree. Next, you should realize that if you are given no information about an individual, you need to assume that the individual is phenotypically normal.

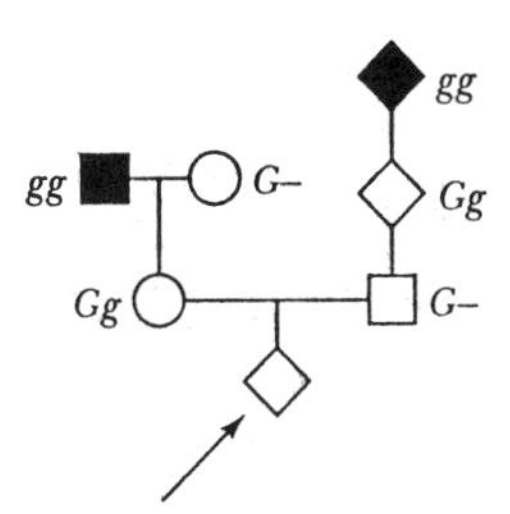

The woman must be *Gg* (*G* from mother, *g* from father), and the chance that she will pass *g* to her child is 1/2. The grandfather had to pass the *g* allele to his child, who must be *Gg*. The chance that the fiance inherited the *g* allele from his parent is 1/2. The chance he will pass it to his child is 1/2. The final probability that the child will be galactosemic is the product of the three probabilities: (1/2)(1/2)(1/2) = 1/8.

12. a. Because Huntington's chorea is rare, the man's father was probably a heterozygote. Therefore, the man has a 50% chance of inheriting the dominant allele.

b. There are two factors that must be considered:

1. probability that the man is *Aa* = 1/2

2. probability that the man, if he is *Aa*, will pass the *A* allele to his child = 1/2

The joint probability (an "and" statement) is (1/2)(1/2) = 1/4.

13. Phenotypically identical parents produced two phenotypes in their offspring. Therefore, they are heterozygous.

a. Dominant. The parents are heterozygotes.

b. $Aa \times Aa \rightarrow 1\ AA : 2\ Aa : 1\ aa$

c. 1/4 probability of normal

3/4 probability of dwarf

14. Both twins could be albino or both twins could be normal ("and" "or" "and" = multiply add multiply). If the parents are heterozygous, both are *Aa*. The probability of being normal (*A-*) is 3/4, and the probability of being albino (*aa*) is 1/4.

p(both normal) + p(both albino)

= p(first normal)p(second normal) + p(first albino)p(second albino)

$= (3/4)(3/4) + (1/4)(1/4)$

$= 9/16 + 1/16 = 5/8.$

15. The plants are 3 spotted : 1 unspotted.

a. Let A = spotted, a = unspotted.

P Aa (spotted) × Aa (spotted)

F_1 1 AA : 2 Aa : 1 aa

3 A- (spotted) *:* 1 aa (unspotted)

b. All unspotted plants should be pure-breeding and one-third of the spotted plants should be pure-breeding.

16.

Cross	Progeny	Conclusion
pin #1 × pin #2	37 pin	One plant is homozygous
thrum #3 × thrum #3	28 thrum	Thrum #3 is homozygous
thrum #3 × pin #1	29 thrum	Thrum is dominant. Thrum #3 is homozygous. Pin #1 and #2 are homozygous recessive.
thrum #4 × pin #2	19 pin 16 thrum	Thrum #4 is heterozygous

a. (1) hh; (2) hh; (3) HH; (4) Hh.

b. (1) $HH \times Hh \rightarrow$ all H-, thrum

(2) $HH \times hh \rightarrow$ all Hh, thrum

(3) $Hh \times Hh \rightarrow$ 1 HH : 2 Hh : 1 hh; 3 thrum : 1 pin

17. In theory, it cannot be proved that an animal is not a carrier for a recessive allele. However, in an A- × aa cross, the more dominant phenotype progeny produced, the less likely it is that one parent is Aa. In such a cross one-half of the progeny would be aa and one-half would be Aa if the parent were Aa. With n dominant phenotype progeny, the probability that the parent is Aa is $(1/2)^n$.

18. Possible reasons for yellow seed:

i. human error

ii. seed color may be determined by maternal genotype (Chapter 20)

iii. yellow may be dominant but not fully penetrant (Chapter 4)

iv. mutation (probably the least likely)

v. nutritional effect

19. The results suggest that winged (*A*-) is dominant to wingless (*aa*) (Cross 2 gives a 3:1 ratio). The five unusual plants are most likely due either to human error or to contamination. Alternatively, they could result from environmental effects on development. For example, too little water may have prevented the seedpods from becoming winged even though they are genetically winged.

20. a. *Pedigree 1*: The best answer is recessive because the disorder skips generations and appears in a mating between two related individuals.

Pedigree 2: The best answer is dominant because it appears in each generation, roughly one-half of the progeny are affected, and affected individuals have an affected parent.

Pedigree 3: The best answer is dominant for the reasons stated for pedigree 2.

Pedigree 4: The best answer is recessive even though it appears in each generation. Two unaffected individuals had one-fourth affected progeny, and the affected individuals in the third generation had an affected father and a mother who could be a carrier.

b. *Genotypes of pedigree 1:*

Generation 1: *AA, aa*

Generation 2: *Aa, Aa, Aa, A-, A-, Aa*

Generation 3: *Aa, Aa*

Generation 4: *aa*

Genotypes of pedigree 2:

Generation 1: *Aa, aa, Aa, aa*

Generation 2: *aa, aa, Aa, Aa, aa, aa, Aa, Aa, aa*

Generation 3: *aa, aa, aa, aa, aa, A-, A-, A-, Aa, aa*

Generation 4: *aa, aa, aa*

Genotypes of pedigree 3:

Generation 1: *Aa, aa*

Generation 2: *Aa, aa, aa, Aa*

Generation 3: *aa, Aa, aa, aa, Aa, aa*

Generation 4: *aa, Aa, Aa, Aa, aa, aa*

Genotypes of pedigree 4:

Generation 1: *aa, A-, Aa, Aa*

Generation 2: *Aa, Aa, Aa, aa, A-, aa, A-, A-, A-, A-, A-*

Generation 3: *Aa, aa, Aa, Aa, aa, Aa*

21.

1/2 *A* — 1/2 B = 1/4 *A B*
1/2 *A* — 1/2 b = 1/4 *A b*

1/2 *a* — 1/2 B = 1/4 *a B*
1/2 *a* — 1/2 b = 1/4 *a b*

22. Each gene should be handled separately. Because for each gene the cross is heterozygote × heterozygote, the expected ratio for each gene is 1:2:1. (This problem is *MM* and *nn* and *Oo*.)

$$p(MM\ nn\ Oo) = p(MM) \times p(nn) \times p(Oo) = (1/4)(1/4)(1/2) = 1/32.$$

23. The cross is

P *AA BB* × *aa bb*

F_1 *Aa Bb*

Half of the alleles from each grandparent can be satisfied by

AA bb = (1/4)(1/4) = 1/16

or +

aa BB = (1/4)(1/4) = 1/16

or +

Aa Bb = (1/2)(1/2) = <u>1/4</u>

3/8

All alleles from one grandparent can be satisfied by

AA BB = (1/4)(1/4) = 1/16

or +

aa bb = (1/4)(1/4) = <u>1/16</u>

1/8

24. a. The ratio of dark : albino and short : long is 3:1. Therefore, each gene is heterozygous in the parents. The cross is *Cc Ss* × *Cc Ss*.

b. Because all progeny are dark, one of the parents is *CC*. The other is *C-*. The ratio of short : long is 1:1, a testcross.

The cross is *C- Ss* : *C- ss*, with one of the parents *CC*. Assuming homozygosity, the cross is *CC Ss* × *CC ss*.

c. All progeny are short (*S-*) and the ratio of dark to albino is 1:1, indicating a testcross. Therefore, the cross is

Cc S- × *cc S-*, with one of the parents *SS*. Assuming homozygosity, the cross is *Cc SS* × *cc SS*.

d. All progeny are albino (*cc*) and the ratio of short to long is 3:1, indicating a heterozygous × heterozygous cross. Therefore, the cross is *ccSs* × *ccSs*.

e. The dark : albino ratio is 3:1, indicating a *Cc* × *Cc* cross. All animals are long (*ss*). The cross is *Cc ss* × *Cc ss*.

f. All animals are dark, indicating at least one parent is homozygous for color. Short : long = 3:1, indicating *Ss* × *Ss*.

The cross is *C- Ss* × *C- Ss*, with one parent *CC*. Assuming homozygosity, the cross is *CC Ss* × *CC Ss*.

g. Dark : albino = 3:1. Short : long = 1:1, a testcross. The cross is *Cc Ss* × *Cc ss*.

25. a. From Cross 3, purple is dominant to green. From Cross 4, cut is dominant to potato.

b. Let *A/a* stand for color and *B/b* stand for shape.

Cross 1: Immediately, the cross can be written *A- B-* × *aa B-*. A 1:1 ratio for color indicates a testcross, and a 3:1 ratio for shape indicates a heterozygous cross. The cross is *Aa Bb* × *aa Bb*.

Cross 2: Immediately the cross can be written *A- B-* × *A- bb*. A 3:1 ratio for color and a 1:1 ratio for shape exists. The cross is *Aa Bb* × *Aa bb*.

Cross 3: Because no green plants exist, the purple parent is homozygous. There is a 3:1 ratio for shape. The cross is *AA Bb* × *aa Bb*.

Cross 4: No potato is seen, therefore the cut is homozygous. There is a 1:1 ratio for color, a testcross. The cross is *Aa BB* × *aa bb*.

Cross 5: A 1:1:1:1 ratio indicates a testcross for each gene. The cross is *Aa bb* × *aa Bb*.

26. **a.** Two genes, shape and skin covering.

b. Bow and knock are alleles, hairy and smooth are alleles.

c. From Cross 1, bow is dominant (call it *B*). From Cross 4, hairy is dominant (call it *H*).

d. *Cross 1*: the two parents are *Bb H-*, with at least one *HH*.

Cross 2: the first is *Bb hh*, the second is *bb hh*.

Cross 3: the first is *Bb Hh*, the second is *bb hh*.

Cross 4: the first is *B- Hh*, the second is *B- Hh*, with at least one *BB*. If the above represents only five parents, then four are *Bb HH, Bb hh, Bb Hh, bb hh*. This leaves *BB H-* (Cross 4) and *B- Hh* (Cross 1) to be reduced to one genotype, which must be *BB Hh*.

27. **a.** Look at each gene separately.

(1) *A- B- C- D- E-* = (1/2)(3/4)(1/2)(3/4)(1/2) = 9/128

(2) *aa B- cc D- ee* = (1/2)(3/4)(1/2)(3/4)(1/2) = 9/128

(3) (question 1) + (question 2) = 9/128 + 9/128 = 9/64

(4) 1 - (question 1) - (question 2) = 1 - 9/128 - 9/128 = 55/64

b. (1) *Aa Bb Cc Dd Ee* = (1/2)(1/2)(1/2)(1/2)(1/2) = 1/32

(2) *aa Bb cc Dd ee* = (1/2)(1/2)(1/2)(1/2)(1/2) = 1/32

(3) (question 1) + (question 2) = 1/32 + 1/32 = 1/16

(4) 1 - (question 1) - (question 2) = 1 – 1/32 –1/32 = 15/16

28. **All families must be *Aa Aa* and must already have at least one affected child. However, not all *Aa Aa* families will be detected. The undetected families must be included to obtain the correct probability.**

First child	Second child	Frequency of family	Total number of children	Total affected children	Total normal children
normal	affected	(3/4)(1/4) = 3/16	6	3	3
affected	normal	(1/4)(3/4) = 3/16	6	3	3
affected	affected	(1/4)(1/4) = 1/16	2	2	0
normal	normal	(3/4)(3/4) = 9/16	18	0	18

That is, 7/16 of the total number of families where both parents are carriers will be detected because of the birth of an affected child. The other families, 9/16 of the total, will not be detected because an affected child did not occur within two births.

Among the detected families, there will be a total of 2 × 7 children, or 14 children. Of these 14 children, 8 will be affected. Therefore, 8/14 = 4/7 of the children will be affected.

29. Let *B*- = brachydactylous, *bb* = normal, *T*- = taster, *tt* = nontaster.

a. *Bb Tt* × *bb Tt*

b. The probability of each child being *Bb* is 1/2. For all eight children it is $(1/2)^8 = 1/256$.

c. The probability of each child not being *Bb* is 1/2. For all eight children it is $(1/2)^8 = 1/256$.

d. The probability of each child being *T*- is 3/4. For all eight children it is $(3/4)^8 = 6{,}561/65{,}536 = 0.1001129$, or 1/10.

e. The probability for each child being *tt* is 1/4. For all eight children it is $(1/4)^8 = 1/65{,}536 = 0.0000152$.

f. The probability of each being *B– T–* = (1/2)(3/4) = 3/8. The probability of all being *B– T–* is $(3/8)^8 = 1/2{,}557 = 0.000391$.

g. The probability of each child not being *B– T–* is 1 – (1/2)(3/4) = 1 – 3/8 = 5/8. The probability of none being *B– T–* is $(5/8)^8$ = 390,625/1,6777,216 = 0.023, or 1/43.

h. The probability of at least one child being *Bb T–* is 1 - the probability of none being *Bb T–* = $1 - (5/8)^8 = 1 - 1/43 = 42/43$.

i. The probability of the first child being *Bb tt* is (1/2)(1/4) = 1/8.

j. The probability of the first two children being *B–* is (1/2)(1/2) = 1/4.

k. Use the formula for combinations to solve this problem:

$$\frac{n!}{p!\,q!}(r)^{p}(s)^{q}$$

where *n* is the total number, *p* is the number of one kind, *q* is the number of the alternative, *r* is the probability of p occurring, and *s* is the probability of *q* occurring. The exclamation point indicates factorial. 5! = (5)(4)(3)(2)(1).

The probability of exactly two out of eight Bb children is thus

$$\frac{8!}{2!\,6!}(1/2)^{2}(1/2)^{6} = \frac{8\times 7}{2}(1/2)^{8} = 28/256 = 1/9$$

This problem also can be done without using the formula for combinations. Assume that the first two children are brachydactylous and the next six are nonbrachydactylous. The probability of that specific combination is $(1/2)^8$. There are a total of 28 different combinations (first child with the second or third or fourth, etc.; second child with the third or fourth or fifth, etc.; third child with the fourth or fifth etc.; etc.), each with the same probability. Therefore, the final probability is $28(1/2)^8 = 1/9$.

1. Either *Bb T-* and *bb tt* or *bb tt* and *BbT-* satisfy the question. "And" indicates multiplication (whether stated or implied as in *Bb* and *tt*), "or" indicates addition. Therefore, the answer is

$$[(1/2)(3/4) \times (1/2)(1/4)] + [(1/2)(1/4) \times (1/2)(3/4)] =$$
$$3/8 \times 1/8 + 1/8 \times 3/8 =$$
$$3/64 + 3/64 = 3/32$$

Tips on Problem Solving

Ratios: A 1:2:1 (or 3:1) ratio indicates that one gene is involved (see Problem 3). A 9:3:3:1 (or some modification of it) ratio indicates that two genes are involved (see Problem 24). A testcross results in a 1:1 ratio if the organism being tested is heterozygous and a 1:0 ratio if it is homozygous (see Problem 2).

Pedigrees: Normal parents have affected offspring in recessive disorders (see Problems 4, 20). Normal parents have normal offspring and affected parents have affected offspring in dominant disorders (see Problems 10, 20). If phenotypically identical parents produce progeny with two phenotypes, the parents were both heterozygous (see Problem 13).

Probability: When dealing with two or more independently assorting genes, consider each gene separately (see Problems 22, 29). The formula for combinations is used to determine the probability that *p* number of events will occur out of a total of *n* number of events (see Problem 29).

Self-Test

1. In humans, when the hands are folded and the fingers are interlocked, placement of the left thumb on top is a dominant condition. Suppose that husband, wife, and one of their three children place their left thumb on top. What are the genotypes of all individuals?

2. The ability to roll the tongue is dominant in humans. Among couples who are both heterozygous and have four children, how many of the families would be expected to show the expected phenotypic ratio?

3. The following crosses were done in cats in order to determine the mode of inheritance of the tail. Evaluate each cross separately, in sequence, and then come to the best conclusion as to the mode of inheritance.

Cross	Parental phenotype	Progeny phenotype
1. cat 1 × cat 2	tailed × tailed	all tailed
2. cat 1 × cat 3	tailed × tailed	all tailed
3. cat 4 × cat 5	no tail × no tail	no tail
4. cat 1 × cat 5	tailed × no tail	3 tailed, 3 no tail

4. A woman gave birth to a child with Tay-Sachs disease, a lethal recessive disorder. What does that tell you about the genotypes of her parents? Her grandparents? Her great-grandparents?

5. Genes D/d, E/e and F/f are independently assorting. If both parents are heterozygous for all genes, what is the probability of

a. an egg that is *D E F*?

b. a sperm that is *D e F*?

c. a child that is *DD Ee F-*?

d. a child that is *D- E- ff*?

6. What are the hallmarks of a pedigree showing recessive inheritance? Dominant inheritance?

7. If a person is heterozygous for five different genes, how many different types of eggs can she produce with respect to those genes?

8. You have a brother who has had several operations for a bone dysplasia that is known to be recessive. Your X rays show evidence of the disorder, but you never needed to have an operation. What are the risks to your children?

9. A child is born with a disorder that is known to have both dominant and recessive forms. Also, a poor diet can cause the same disorder. How would you go about determining which form of the disorder the child has?

10. A dog has the dominant phenotype *D-*. How can you determine if it is homozygous or heterozygous?

Solutions to Self-Test

1. Husband and wife: *Rr*. Child with left thumb on top: *R-*. Children with right thumb on top: *rr*.

2. The expected phenotypic ratio is 3:1. Some families would be 3:1, but many would not be. The formula to use is

$$\frac{n!}{p!\,q!}(r)^{p}(s)^{q}$$

where n is the total number, p is the number of one kind, q is the number of the alternative, r is the probability of p occurring and s is the probability of q occurring. The exclamation point indicates factorial. $5! = (5)(4)(3)(2)(1)$.

$$(4!)(3/4)^3(1/4)^1/(3!)(1!) = 4(27/64)(1/4) = 27/64 = 42.2\%$$

In other words, only 42.2% of the families would actually have a 3:1 ratio.

3.

Cross	Conclusion
1	At least one of the parents is homozygous.
2	At least one of the parents is homozygous.
3	At least one of the parents is homozygous.
4	At least one of the parents is heterozygous.

Nothing more than the above can be concluded from these crosses.

4. At least one person in each generation must have been a carrier (heterozygous) for Tay-Sachs disease.

Parents: 1 in 2

Grandparents: 1 in 4

Great grandparents: 1 in 8

5. a. $(1/2)(1/2)(1/2) = 1/8$

b. $(1/2)(1/2)(1/2) = 1/8$

c. $(1/4)(1/2)(3/4) = 3/32$

d. $(3/4)(3/4)(1/4) = 9/64$

6. *Recessive:* Affected people may have normal parents; unaffected people may have an affected child; two affected people have only affected children; among the people who are genetically related, one-fourth may be affected.

Dominant: Affected people have at least one affected parent; two unaffected people have unaffected children; among the people who are genetically related, one-half may be affected.

7. For each gene there are two alternatives. Therefore, there are 2^5 different eggs that she may produce.

8. Not all people who have the genotype for a disorder necessarily have the phenotype associated with it. You have some aspects of the phenotype at the X ray level, which means that you also have the genotype for the disorder. However, unless you marry someone who is a carrier, your children should be symptom-free, although one-half would be expected to be carriers.

9. Assuming that testing cannot distinguish among the three possible causes, the only way to pinpoint the cause is to do a complete family history that would include questions about diet.

10. A testcross with *dd* will result in all *D-* progeny if the dog is *DD* and in 1 *D-* : 1 *dd* if the dog is heterozygous *Dd*.

CHAPTER

3

Chromosome Theory of Inheritance

Important Terms and Concepts

The **chromosome theory of heredity** means that genes are located on chromosomes.

Mitosis is the orderly distribution of one chromatid from each chromosome to each of two daughter cells. The chromosome number remains constant between cell generations. The stages of the mitotic cycle are interphase, prophase, metaphase, anaphase, and telophase.

Meiosis consists of two sequential cell divisions. Usually meiosis I (reductional division) is the orderly distribution of one chromosome from each chromosome pair to each of two daughter cells. The chromosome number is reduced by half. The stage between meiosis I and II is called interkinesis. Usually meiosis II (equational division) is the orderly distribution of one chromatid from each chromosome to each of two daughter cells. In form, it is identical to mitosis. The chromosome number remains constant between cell generations. Both meiotic divisions are subdivided into a number of stages.

The **nuclear spindle** is a protein structure that aids in proper chromosome movement during mitosis and meiosis.

Homologs (also spelled **homologues**) are two chromosomes that carry genes for the same functions.

Synapsis is the pairing of homologous chromosomes prior to meiosis. The process occurs with the aid of the synaptinemal complex, a protein structure.

A **genome** (usually one *n*) consists of one set of unique chromosomes.

A **diploid** (2*n*) cell or organism contains two homologous genomes, one from each parent.

A **haploid** (1*n*) cell or organism has one genome.

A **chiasma** is the chromosomal structure that is taken as the visible proof of the molecular event of crossing-over.

A **heteromorphic pair** is a pair of sex chromosomes that are nonidentical in shape and/or size and presumably have only partial homology. Genes on these chromosomes are called **hemizygous.**

X-linkage refers to genes located on the X chromosome.

Autosomes are the paired (in diploids) nonsex chromosomes of both sexes.

Y-linkage refers to genes located on the Y chromosome.

The number of sets of genomes present in an organism varies with the species, as does which sex is the heterogametic sex (if sexual differentiation exists). Some organisms are mostly diploid, some mostly haploid, and some alternate between haploid and diploid. Other organisms may routinely have more than two genomes (see Chapter 9). In all species that have meiosis, the laws of Mendel apply.

Be sure that you have thoroughly read the entire chapter before you attempt any of the problems.

Solutions To Problems

1. **i.** In mitosis the chromosome number remains unchanged while in meiosis the chromosome number is halved.

ii. In mitosis sister chromatids separate from each other while in meiosis both homologous chromosomes and sister chromatids separate from each other.

iii. Mitosis leads to two cells while meiosis leads to four cells.

iv. Homologous pairing occurs only in meiosis.

v. Recombination is much more frequent in meiosis than mitosis.

2. Because mitosis does not involve the separation of allelic alternatives, the daughter cells will both be *Aa Bb Cc*.

3. P $ad^- \, a \times ad^+ \, \alpha$

Transient diploid $ad^+/ad^- \; a/\alpha$

F_1 1 $ad^+ \, a$

1 $ad^- \, a$

1 $ad^+ \, \alpha$

1 $ad^- \, \alpha$

4. P $s^+/s^+ \times s/Y$

F_1	1 s^+/s	normal female
	1 s^+/Y	normal male
F_2	1 s^+/s^+	normal female
	1 s^+/s	normal female
	1 s^+/Y	normal male
	1 s/Y	small male

In the cross $s^+/s \times s/Y$, the progeny will be

1 s^+/s	normal female
1 s/s	small female
1 s^+/Y	normal male
1 s/Y	small male

5.

	Mitosis	Meiosis
fern	sporophyte gametophyte	prothallus
moss	sporophyte	archegonia

	gametophyte	antheridia
flowering plant	sporophyte	flowers
	gametophyte	
pine tree	sporophyte	pinecones
	gametophyte	
mushroom	sporophyte	hyphae
	gametophyte	
frog	somatic cells	gonads
butterfly	somatic cells	gonads
snail	somatic cells	gonads

6. Using the Mendelian convention, B = brown and b = black. Using *Drosophila* symbolism, B = brown and B^+ = black. If you wanted to convey more information with your symbols, you could use Br for brown and either br or Br^+ for black.

7. This problem is tricky because the answers depend on how a cell is defined.

a. 46 chromosomes, each with 2 chromatids = 92 chromatids

b. 46 chromosomes, each with 2 chromatids = 92 chromatids

c. 46 chromosomes in each of 2 about-to-be-formed cells, each with 1 chromatid

d. 23 chromosomes in each of 2 about-to-be-formed cells, each with 2 chromatids,

e. 23 chromosomes in each of 2 about-to-be-formed cells, each with 1 chromatid

8. e. chromosome pairing

9. a. There is no reduction in the number of chromosomes during mitosis, just a reduction in the number of chromatids/chromosome. Therefore, the answer is (4) 20 (or 10 pairs).

b. Meiosis involves the reduction of the number of chromosomes and the number of chromatids/chromosome by a factor of 2 each. Therefore, the cell goes from 10 pairs of chromosomes, each with 2 chromatids, to 10 chromosomes, each with 1 chromatid.

10. His children will have to inherit the satellite 4 (probability = 1/2), the abnormally staining 7 (probability = 1/2), and the Y chromosome (probability = 1/2). To get all three, the probability is (1/2)(1/2)(1/2) = 1/8.

11. The parental set of centromeres can match either parent, which means there are two ways to satisfy the problem. For any one pair, the probability of a centromere from one parent going into a specific gamete is $1/2$. For n pairs, the probability of all the centromeres being from one parent is $(1/2)^n$. Therefore, the total probability of having a haploid complement of centromeres from either parent is $2(1/2)^n = (1/2)^{n-1}$.

12. a. Your mother gives you 23 chromosomes, one-half of all that she has and one-half of all that you have. Therefore, you have one-half of all your genes in common with your mother.

b. For any heterozygous gene (*Aa*) in your mother,

You	Brother
A	*A*
A	*a*
a	*A*
a	*a*

Thus, there are two of four combinations between you and you brother that will be a match. The same holds for any heterozygous gene in your father. Therefore, you and your brother will have one-half of your genes in common for each gene that is heterozygous in your parents. You and your brother will have identical alleles from your parents for each gene that is homozygous in your parents. The degree of homozygosity in a parent is unknown. For this reason, the final answer can be stated only for heterozygosity in the parents, and homozygosity must be ignored.

13. Meiosis will have to result in one X per gamete in the female and one X or two Y's in the male. Any scheme that results in this is sufficient.

14. First assume that the cross is autosomal:

P *G-* (graceful female) × *gg* (gruesome male)

F_1 1 *Gg* graceful female

1 *g-* female

1 *Gg* graceful male

1 *g-* male

This cross does not meet the observation, so it must be wrong. The cross also cannot be *G-* X *g*Y (X-linked) because graceful females will result.

Next assume that the female is the heterogametic sex.

P *GO* (graceful female) × *gg* (gruesome male)

F_1 1 *gO* gruesome female

1 *Gg* graceful male

The O can be a female-determining chromosome or no chromosome.

15. The disease cannot be autosomal because there is a sexual split in the progeny. It cannot be X-linked recessive for two reasons: (1) males get their Y from their fathers and couldn't have the disease, and (2) females get a normal X from their mother, who is unlikely to be a carrier. If the mother *is* a carrier, it is highly unlikely she would pass the gene to all of her daughters and none of her sons. The daughters will show any dominant X-linked gene that their father has because he must pass it to them. The answer is **d.**, X-linked dominance.

16. Let H = hypophosphatemia and h = normal. The cross is $HY \times hh$ $\rightarrow Hh$ (females) and hY (males). The answer is **e.**, 0.

17. *If* the historical record is accurate, the data suggest Y-linkage. Another explanation is an autosomal gene that is dominant in males and recessive in females. This has been observed for other genes in both humans and other species.

18. You should draw pedigrees for the following problems.

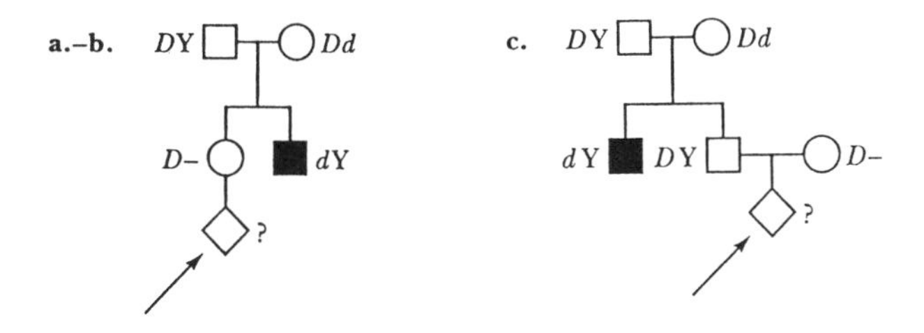

a. The probability that the woman inherited the d allele from her mother is 1/2. The probability that she passes it to her child is 1/2. The probability that the child is male is 1/2. The total probability of the woman having an affected child is (1/2)(1/2)(1/2) = 1/8.

b. Your maternal grandmother had to be a carrier, Dd. The probability that your mother received the gene is 1/2. The probability that your mother passed it to you is 1/2. The total probability is (1/2)(1/2) = 1/4.

c. Because your father does not have the disease, you cannot inherit it from him. The probability is 0.

19. a. Because neither parent shows it, the disease must be recessive. Because of the sexual split in the progeny, it is most likely X-linked. If it were autosomal, all three parents would have to carry it, which is rather unlikely.

b. P AY, Aa, AY

F_1 AY, A-, aY, A-, AY, aY, aY, A-, aY, A-

20. You should draw the pedigree before beginning.

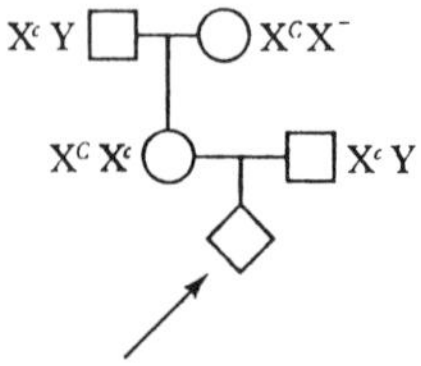

X^cY — X^CX^-

X^CX^c — X^cY

a. Cc, cc

b. p(color blind) p(male) = (1/2)(1/2) = 1/4.

c. The girls will be 1 normal (Cc) : 1 color blind (cc)

d. The cross is $Cc \times cY \rightarrow$ 1 normal : 1 color blind for both sexes.

21. a. The complete answer to this problem involves X-inactivation (Chapter 14). For now, the colors can be explained by assuming that both alleles are expressed in the female. Let B = black and Y = orange.

Females	Males
X^BX^B = black	X^BY = black
X^YX^Y = orange	X^YY = orange
X^BX^Y = tortoise	

b.

P	X^YX^Y	(orange) × X^BY (black)
F_1	X^BX^Y	tortoise-shell female
	X^YY	orange male

c.

P	X^BX^B	(black) × X^YY (orange)
F_1	X^BX^Y	tortoise female
	X^BY	black male

d. Because the males are black or orange, the mother had to have been tortoise-shell. One-half the daughters are black, indicating homozygosity, which means that their father was black.

e. Males were orange or black, indicating that the mothers were tortoise-shell. Orange females, indicating homozygosity, mean that the father was orange.

22. e. 1/4

23. c. X-linked recessive

24. a. X-linked recessive

b. *Generation 1*: X^+/Y, X^+/X^x

Generation 2: X^+-, X^x/Y, X^+/Y, X^+_, X^+/X^x, X^+/Y

Generation 3: X^+–, X^+/Y, X^+/X^x, X^+/X^x, X^+/Y, X^+–, X^x/Y, X^+/Y, X^+-

c. The first couple has no chance of an affected child because the son received his Y chromosome from his father. The second couple has a 50 percent chance of having affected sons and no chance of having affected daughters. The third couple has no chance of having an affected child.

25. Possible responses include orderly division at mitosis and meiosis, more efficient gene regulation or function, and maintenance of favorable gene combinations.

26. a. autosomal recessive: excluded by unaffected female in third generation

b. autosomal dominant: consistent

c. X-linked recessive: excluded by affected female with unaffected father

d. X-linked dominant: excluded by unaffected female in third generation

e. Y-linked: excluded by affected females

27. a. Cross 6, bent × bent ---> normal, indicates that bent is dominant.

b. The sexual differences in phenotype in cross 6 indicate that it is X-linked.

c. Let B = bent and b = normal.

	Parents		Progeny	
Cross	Female	Male	Female	Male
1	*bb*	*B*Y	*Bb*	*b*Y
2	*Bb*	*b*Y	*Bb*, *bb*	*B*Y, *b*Y
3	*BB*	*b*Y	*Bb*	*B*Y

4	*bb*	*b*Y	*bb*	*b*Y
5	*BB*	*B*Y	*BB*	*B*Y
6	*Bb*	*B*Y	*BB*, *Bb*	*B*Y, *b*Y

28. a. An extra finger is autosomal dominant because all affected individuals have an affected parent, it occurs in every generation, and male-to-male transmission occurs.

b. Eye disease is most likely X-linked recessive because only males have it and affected males have unaffected parents. If it were autosomal recessive, the father of individual 11 would have to be a carrier in addition to the mother being a carrier.

c. Let F = extra finger and E = normal eyes. The cross is *Ee ff* (female) X *EY Ff* (male).

d. Individual 4 does not have an extra finger so her child cannot inherit it. Because her father does not have the eye disease, she also does not have the gene for it.

e. Individual 12 does not have an extra finger; her child cannot have it. She could be a carrier (50 percent chance) of the eye disease. If so, the child has a 50 percent chance of inheriting the gene and a 50 percent chance of being a male, or (1/2)(1/2)(1/2) = 1/8 probability of having the eye disease.

29. a.

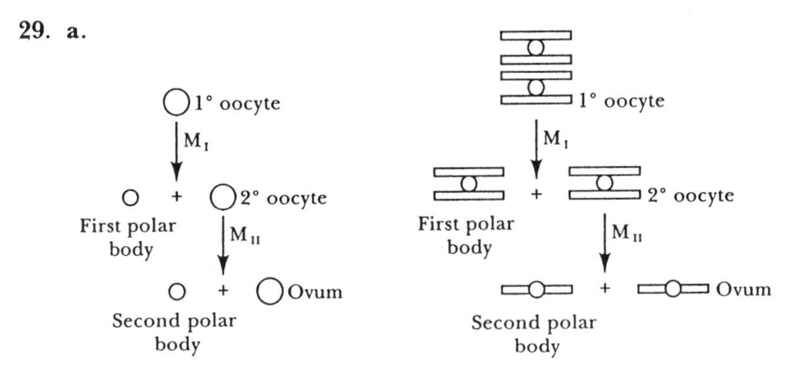

b. False. Segregation of any gene pair is random and the probability of any one allele being included in an ovum is 1/2. In a large number of ova produced by a given female, there is a 1:1 probability of inclusion for each member of an allelic pair.

30. a.

$1/4\ p^+ h^+$ $\quad 1/4\ p\ h^+$

$1/4\ p^+ h$ $\quad 1/4\ p\ h$

b. Because both genes are expressed in the sporophyte, they would not necessarily be expressed in the gametophyte.

c. This is a dihybrid cross and the expected result are

9 $p^+ h^+$	green, hairy	3 $p\ h^+$	pink, hairy
3 $p^+ h$	green, hairless	1 $p\ h$	pink, hairless

31. Start this problem by writing the crosses and results so that all the details are clear.

P brown, short female × red, long male

F_1 red, long females

red, short males

These results tell you that long is dominant to short and that the chromosome carrying the gene is X-linked. The results also tell you that eye color is autosomal and that red is dominant to brown. Let b = brown, B = red, s = short, and S = long. The cross can be rewritten as follows:

P $b/b\ s/s \times B/B\ S/\mathrm{Y}$

F_1 $B/b\ S/s$ females

$B/b\ s/\mathrm{Y}$ males

F_1 gametes

female: 1 BS : 1 Bs : 1 bS : 1 bs

male: 1 Bs : 1 bs : 1 $B\mathrm{Y}$: 1 $b\mathrm{Y}$

F_2 females

1/16 $B/B\ S/s$ red, long

1/16 $B/B\ s/s$ red, short

2/16 $B/b\ S/s$ red, long

2/16 $B/b\ s/s$ red, short

1/16 $b/b\ S/s$ brown, long

1/16 $b/b\ s/s$ brown, short

males

1/16 $B/B\ S/\mathrm{Y}$ red, long

1/16 $B/B\ s/\mathrm{Y}$ red, short

2/16 $B/b\ S/\mathrm{Y}$ red, long

2/16 $B/b\ s/Y$ red, short

1/16 $b/b\ S/Y$ brown, long

1/16 $b/b\ s/Y$ brown, short

The final phenotypic ratio is

3/8 red, long	1/8 brown, short
3/8 red, short	1/8 brown, long

32. **a** and **b**. In this particular pedigree, the second generation shows 50 percent affected. In the third generation, offspring of affected individuals show a very high rate of the disorder, while offspring of nonaffected individuals are not affected. This is a typical dominant pattern.

In the sixth generation, the sole affected male has an affected father, indicating autosomal location.

Inbreeding exists in the third generation, and inbreeding always raises the question of recessive inheritance. While a recessive pattern of inheritance is compatible with the entire pedigree, the pattern of disease in the second generation suggests dominance. An assumption of recessive inheritance would require that three individuals from the general population carry the recessive allele. Because this is unlikely, the best interpretation is autosomal dominant.

c. Both male and female offspring would have a 50 percent risk of being affected.

Tips on Problem Solving

Problem 14 clearly illustrates the process of successive generation of hypotheses, which are then tested against the data.

X-linkage or autosomal: if the male phenotype is different from the female phenotype, X-linkage is involved for the allele carried by the female (see Problems 15, 27).

Inheritance patterns: There are only four possible inheritance patterns for a gene (see Problem 26):

1. autosomal dominant
2. autosomal recessive
3. X-linked dominant
4. X-linked recessive

Self-Test

1. When does Mendel's first law occur?

2. When does Mendel's second law occur?

3. How many DNA molecules would there be per cell in a mature sperm from an organism in which $2n = 18$?

4. How many chromatids per cell would there be in the first polar body of a mammal with $2n = 24$?

5. In a haploid organism in which $1n = 15$, how many chromosomes would there be in a cell beginning meiosis I?

6. Outline a decision-making flowchart for deciding on the mode of inheritance from data in a human pedigree.

7. In some organisms, the male is the homogametic sex. For a recessive sex chromosome-linked disorder, which sex would be affected at a higher rate?

8. Two heterozygotes for independently assorting genes *A/a* and *B/b* mate. Among their offspring, there are only two phenotypic classes, *A- B-* and *aa bb*, in a 9:7 ratio. How can you account for this?

9. If a person is *Dd E/Y*, what gametes will he produce?

10. Genes *D/d*, *E/e* and *F/f* are independent. What will be the progeny and their ratios in a mating of the following individuals: *Dd EE Ff* × *Dd ee FY*?

Solutions to Self-Test

1. The first law states that alleles segregate from each other, which occurs at the onset of anaphase I.

2. The second law states that genes independently assort, which occurs at the onset of anaphase I.

3. If $2n = 18$, a mature sperm would have nine chromosomes, each with one chromatid. There would be one DNA molecule per chromosome, or nine DNA molecules.

4. The first polar body in mammals contains $1n$ of chromosomes, each with two chromatids. If $2n = 24$, $1n = 12$. Therefore, there would be 24 chromatids in the polar body.

5. To be entering meiosis I, fertilization or fusion must have occurred, so there would be 30 chromosomes.

6. There are two decisions that need to be made: (1) whether the disorder is dominant or recessive and (2) whether it is X-linked or autosomal. The decisions can be made in either order, but it is often easier to decide between X-linked and autosomal first.

1. Is the disorder X-linked or autosomal?

a. What is the sex ratio among affected individuals? If equal, then may be autosomal. If clearly skewed, X-linked.

b. Is there male to male transmission? If yes, then autosomal. If no, may be X-linked.

2. If the disorder is X-linked:

a. If females are affected at a greater rate than males, the disorder is dominant.

b. If males are affected at a greater rate than females, the disorder is recessive.

3. If the disorder is autosomal:

a. Do affected individuals have affected parents? If yes, possibly dominant. If no, recessive.

b. Do normal parents have affected offspring? If yes, recessive. If no, dominant.

7. females

8. The two genes somehow affect the expression of each other. *A- B-* is one phenotypic class; all other gene combinations constitute the other phenotypic class.

9. *DE, dE, DY, dY*

10.

1/16 *DD Ee FF*	1/8 *Dd Ee FF*	1/16 *dd Ee FF*
1/16 *DD Ee FY*	1/8 *Dd Ee FY*	1/16 *dd Ee FY*
1/16 *DD Ee Ff*	1/8 *Dd Ee Ff*	1/16 *dd Ee Ff*
1/16 *DD Ee fY*	1/8 *Dd Ee fY*	1/16 *dd Ee fY*

CHAPTER

4

Extensions to Mendelian Analysis

Important Terms and Concepts

Incomplete dominance produces an intermediate phenotype in the heterozygote. A 1:2:1 ratio is observed. Examples are flower color (red and white → pink) and enzyme activity (high and low → medium).

Codominance is revealed when the heterozygote possesses the phenotype of both homozygotes. A 1:2:1 ratio is observed. Examples are hemoglobin variants (Hb^A and $Hb^S \rightarrow Hb^A + Hb^S$) and phosphoglucomutase variants (PGM-1 and PGM-2 → PGM-1 + PGM-2) as determined by electrophoresis.

For allelic interactions, what you determine to be the mode of interaction depends quite frequently on the manner in which the trait is observed. Consider the heterozygote Hb^AHb^S. If you look at the red blood cells under normal conditions, you conclude that Hb^A is dominant to Hb^S.

Under conditions of low oxygen tension the cells sickle, leading to the opposite conclusion. At the level of electrophoresis of the globin proteins, however, the conclusion is codominance.

Multiple alleles lead to single-gene-pair ratios (1:1, 3:1, 1:2:1).

Multiple genes lead to ratios indicating two genes (9:3:3:1) or three genes (27:9:9:9:3:3:3:1), or some modification of the ratios, in heterozygote × heterozygote crosses.

Recessive lethal genes cause a distortion in the expected ratio. An example is a 2:1 in a heterozygote X heterozygote cross.

Epistasis is the alteration of expression of one gene by the expression of another gene. Distortions of expected ratios are observed.

Modifier genes modify the expression of other genes, for example light and dark color. They are detected in heterozygote X heterozygote crosses by a 9:3:3:1 ratio.

Complementary genes work together to produce a phenotype. Heterozygote X heterozygote crosses usually yield a 9:7 ratio.

Suppression occurs when one gene blocks the expression of another. In F_1 crosses, the ratios can be 13:3, 12:4, or some other variant on the expected ratio.

Penetrance is a populational term and is defined as the percentage of individuals of a given genotype that express the phenotype associated with the genotype. A lack of penetrance in an individual is due to epistatic relationships between genes. The environment may also be a factor.

Expressivity is the extent to which a phenotype is expressed in an individual. The range is from minimal to full, depending upon the effects of epistatic genes and environmental factors.

Pleiotropy is the phenomenon of multiple effects from one gene. This results in a syndrome (a collection of symptoms associated with a particular disorder).

Dominance occurs within genes, between alleles, whereas epistasis occurs between genes.

Be sure that you have thoroughly read the entire chapter before you attempt any of the problems.

Solutions to Problems

1. The woman must be *AO*, so the mating is $AO \times AB$. Their children will be

Genotype	Phenotype
1 *AA*	*A*
1 *AB*	*AB*
1 *AO*	*A*
1 *BO*	*B*

2. **a.**

CC II	white	*CC Ii*	white	*CC ii*	colored
Cc II	white	*Cc Ii*	white	*Cc ii*	colored
cc II	white	*cc Ii*	white	*cc ii*	white

b. The cross was *Cc Ii* × *Cc Ii*, yielding the typical

9 *C– I*	white	3 *C– ii*	colored
3 *cc I–*	white	1 *cc ii*	white

3. The hairless dogs are heterozygous (*Hh*), and *H* is a recessive lethal. The cross is *Hh* × *Hh*. The progeny are

1 *HH*	deformed, dead
2 *Hh*	hairless
1 *hh*	normal.

To test this hypothesis, cross a hairless (*Hh*) with a normal (*hh*). The progeny should be 1/2 hairless, 1/2 normal.

4. The first sentence tells you that white is recessive to pink. Therefore, pink flowers will be either *AA* or *Aa*, and white flowers will be *aa*. The cross that is described in the second sentence is *Aa* × *aa*. This is a typical testcross, with an expected phenotypic ratio of 1 pink : 1 white. Therefore, the correct answer is **a**.

5. You are told that the cross of two erminette fowls results in 22 erminette, 14 black, and 12 pure white. Two facts are important: (1). the parents consist of only one phenotype, yet the offspring have three phenotypes; and (2). the progeny appear in an approximate ratio of 1:2:1. These facts should tell you immediately that you are dealing with a heterozygous × heterozygous cross and that the erminette phenotype must be the heterozygous phenotype.

When the heterozygote shows a different phenotype than either of the two homozygotes, the heterozygous phenotype results from incomplete dominance or codominance. Because two of the three phenotypes contain black, either fully or in an occasional feather, you might classify the erminette as an instance of incomplete dominance because it is intermediate between fully black and fully white. Alternatively, because the erminette has both black and white feathers, you might classify the phenotype as codominant. Your decision will rest on whether you look at the whole animal (incomplete dominance) or at individual feathers (codominance). This is yet another instance where what you conclude is determined by how you observe.

To test the hypothesis that the erminette phenotype is a heterozygous phenotype, you could cross an erminette with either, or both, of the homozygotes. You should observe a 1:1 ratio in the progeny of both crosses.

6. The solution to this problem lies in the recognition that the sporophyte is a diploid heterozygote (*P1 P2*). At the level of electrophoresis of proteins, you will observe codominance and, therefore, two proteins in sporophytes. Because the gametophytes are haploid, the spores will have either P1 or P2 and the ratio will be 1:1.

7. P *S1 S3* × *S2 S4*

F_1 *S1 S2, S1 S4, S2 S3, S3 S4*

The following F_1 crosses need to be evaluated:

Female	Male	Number of alleles in common
S1S2	*S1S2*	2
S1S2	*S1S4* + reciprocal	1
S1S2	*S2S3* + reciprocal	1
S1S2	*S3S4* + reciprocal	0
S1S4	*S1S4*	2
S1S4	*S2S3* + reciprocal	0
S1S4	*S3S4* + reciprocal	1
S2S3	*S2S3*	2
S2S3	*S3S4* + reciprocal	1
S3S4	*S3S4*	2

The hard way is to analyze all 16 crosses. The easy way is to recognize the following rules:

1. If the parents are not alike, the cross is fully fertile.
2. If the parents share one allele, the cross is partially fertile.
3. If the parents share two alleles, the cross is fully sterile.
4. Reciprocal crosses will yield the same results.

There are four crosses that involve totally identical alleles, which results in full sterility.

There are eight crosses that involve one identical allele between the parents. This will result in partial fertility.

The remaining four crosses involve nonidentical alleles. This will result in full fertility.

Thus, the final ratio is 1/4 fully fertile : 1/2 partially fertile : 1/4 fully sterile.

8. P *S1 S2* × *S3 S4*

F_1	Genotype	Type
	S1S3	*S1*
	S1S4	*S1*
	S2S3	*S2*
	S2S4	*S2*

Because compatibility is determined by type, the only F_2 crosses that are compatible are *S1* × *S2*.

S1 S3 × *S2 S3* → *S1 S2, S1 S3, S2 S3, S3 S3*

S1 S4 × *S2 S4* → *S1 S2, S1 S4, S2 S4, S4 S4*

S1 S3 × *S2 S4* → *S1 S2, S1 S4, S2 S3, S3 S4*

S1 S4 × *S2 S3* → *S1 S2, S1 S3, S2 S4, S3 S4*

9. The easy way to solve this problem is to note that the Himalayan phenotype is observed only in homozygotes. Because one parent does not have the Himalayan allele, it will be impossible to obtain progeny that are homozygous for it. The answer is, therefore, **e**.

The more scholarly approach to this problem is to look at the progeny and their ratio before concluding that Himalayan will not be observed. The progeny are

1 C^+C^{ch} full color 1 $C^{ch}C^{ch}$ chinchilla

1 C^+C^h full color 1 $C^{ch}C^h$ chinchilla

10. a. You could begin this problem using one of several assumptions. Only one will be right, but there is no way to know in advance which that will be. The correct assumption is that there is one gene, with multiple alleles. Symbolize the different alleles using the first letter of each color.

Cross 1: $ba \times ba \rightarrow 3\ b- : 1\ aa$

Cross 2: $bs \times aa \rightarrow 1\ ba : 1\ sa$

Cross 3: $ca \times ca \rightarrow 3\ c- : 1\ aa$

Cross 4: $sa \times ca \rightarrow 1\ ca : 2\ s- : 1\ aa$

Cross 5: $bc \times aa \rightarrow 1\ ba : 1\ ca$

Cross 6: $bs \times c- \rightarrow 1\ b- : 1\ s-$

Cross 7: $bs \times s- \rightarrow 1\ b- : 1\ s-$

Cross 8: $bc \times sc \rightarrow 2\ b- : 1\ sc : 1\ cc$

Cross 9: $sc \times sc \rightarrow 3\ s- : 1\ cc$

Cross 10: $ca \times aa \rightarrow 1\ ca : 1\ aa$

The order of dominance is $b > s > c > a$.

b. The cross between parents is *bs* X *bc*. The progeny are

1 *bb* black	1 *bs* black
1 *bc* black	1 *sc* sepia

11. a. This suggests that the product of the Himalayan allele is an enzyme that is temperature-sensitive. At the higher core body temperature the enzyme's structure is disrupted and thus the enzyme cannot function, resulting in no color. At the lower temperatures of the extremities, the enzyme is in an active configuration and results in color.

b. Because the area on the back of the mouse is kept cold, the enzyme product of the Himalayan allele is functional. Therefore, that area would be black when the hair grows back.

12. Both codominance and classical dominance are present in the multiple allelic series for blood type: $A = B$, $A > O$, $B > O$.

The *O* baby is homozygous for *O* (*OO*). Thus both parents had to have an *O* allele. Parent sets **b** ($A \times O$) and **d** ($O \times O$) fulfill this requirement. Formally, no decision can be made between the two until the other babies are considered, but it should be obvious that the $O \times O$ parents can have only an *O* child.

The *A* baby is either homozygous for *A* (*AA*) or is heterozygous (*AO*). Parent sets **a** ($AB \times O$), **b** ($A \times O$), and **c** ($A \times AB$) could be the parents. No decision can yet be made.

The *B* baby is either homozygous (*BB*) or heterozygous (*BO*). Only parent set **a** (AB × *O*) could result in a *B* child. Therefore, the *B* child is genotypically *BO*.

The *AB* baby must come from a cross where one parent has an *A* (*A*– or *AB* phenotype) and one parent has a *B* (*B*– or *AB* phenotype). Only parent set **c** ($A \times AB$) meets this requirement.

Because parents **c** must be the parents of the *AB* baby, they cannot be the parents of the *A* baby. That leaves only parents **b** for baby *A*, and baby *A* must be *AO* genotypically. If parents **b** must be the parents of baby *A*, they cannot be the parents for baby *O*. Therefore, only parents **d** can be the parents for baby *O*.

13. *M* and *N* are codominant alleles. The rhesus group is determined by classically dominant alleles. The *ABO* alleles are mixed codominance and classical dominance (see Problem 12).

The mother had to donate an *N* to her children. Because one of the children is *rr*, the mother is *Rr* and could donate either allele to her children. For similar reasons, the mother must be *AO* and could donate either allele.

Child 1 is homozygous for *O*, and he had to obtain an *O* allele from his father. Because the lover is *AB*, he cannot be the father. The husband could be the father because he has two *O* alleles. He also could have donated *M* and either *R* or *r* (assuming he is *Rr*) to the child.

Child 2 is homozygous for *N* and had to obtain an *N* allele from his father. Because the husband is homozygous for *M*, he cannot be the father. The lover, who is *MN*, could be the father. He also could have donated *A* and *r* to the child.

Child 3 is either *AO* or *AA*, is *MN*, and is homozygous for Rh^- (*rr*). With respect to *ABO*, either man could be the father because the husband could have donated an *O* and the lover could have donated an *A*. With respect to *MN*, both men have the required *M* allele (the *N* had to have come from the mother). With respect to Rh^-, the husband could be *Rr* and the lover is *rr*. Therefore, either man could be the father of this child.

14. The key to solving this problem is in the statement that breeders cannot develop a pure-breeding stock and that a cross of two platinum foxes results in some normal progeny. Platinum must be dominant to normal color and heterozygous (*Aa*). An 82:38 ratio is very close to a 2:1. Because a 1:2:1 ratio is expected in a heterozygous cross, one genotype is nonviable. It must be the *AA*, homozygous platinum, genotype that is nonviable because the homozygous recessive genotype is normal color (*aa*). Therefore, the platinum allele is a pleiotropic allele that governs coat color in the heterozygous state and is lethal in the recessive state.

15. a. Because Pelger crossed with normal results in two phenotypes in a 1:1 ratio, either Pelger or normal is heterozygous (*Aa*) and the other is homozygous (*aa*) recessive. The problem states that normal is true breeding, or, *aa*. Pelger must be *Aa*.

b. The cross of two Pelger rabbits results in three phenotypes. This means that the Pelger anomaly is dominant to normal. This cross is $Aa \times Aa$, with an expected ratio of 1:2:1. Because the normal must be aa, the extremely abnormal progeny must be AA. There were only 39 extremely abnormal progeny because the others died before birth.

c. The Pelger allele is pleiotropic. In the heterozygous state it is dominant for nuclear segmentation of white blood cells. In the homozygous state it is a recessive lethal.

You could look for the nonsurviving fetuses in utero. Because the hypothesis of embryonic death of the homozygous dominant predicts a one-fourth reduction in litter size, you could also do an extensive statistical analysis of litter size, comparing normal × normal with Pelger × Pelger.

d. By analogy with rabbits, the absence of a homozygous Pelger anomaly in humans can be explained as recessive lethality. Alternatively, because 1 in 1000 births results in a Pelger anomaly, a heterozygous × heterozygous mating would be expected in only 1 of 1 million (1/1000 × 1/1000) matings, and only 1 in 4 of the progeny would be expected to be homozygous. Thus, the homozygous Pelger anomaly is expected in only 1 of 4 million births. This is extremely rare and might not be recognized.

e. By analogy with rabbits, the child who is homozygous for the Pelger allele would be expected to have severe skeletal defects.

16. Note that a cross of the short-bristle female with a normal male results in two phenotypes with regard to bristles and an abnormal sex ratio of 2 females : 1 male. Furthermore, all the males are normal, while the females are normal and short in equal numbers. Whenever the sexes differ with respect to phenotype among the progeny, an X-linked gene is involved. Because only the normal phenotype is observed in males, the short bristle phenotype must be heterozygous, and the allele must be a recessive lethal. Thus the first cross was $Aa \times aY$.

Long-bristle females (aa) were crossed with long-bristle males (aY). All their progeny would be expected to be long-bristle (aa or aY).

Short-bristle females (Aa) were crossed with long-bristle males (aY). The progeny expected are

1 Aa short females	1 aY long males
1 aa long females	1 AY nonviable

17. In order to do this problem, you need first to restate the information provided. The following two genes are independently assorting:

hh = hairy ss = no effect

Hh = hairless *SS* = lethal

HH = lethal *Ss* suppresses *Hh*, giving hairy

a. The cross is *Hh Ss* × *Hh Ss*. Because this is a typical dihybrid cross, the expected ratio is 9:3:3:1. However, the problem cannot be worked in this simple fashion because of the epistatic relationship of these two genes. Therefore, the following approach should be used.

For the *H* gene, you expect 1/4 *HH* : 1/2 *Hh* : 1/4 *hh*. For the *S* gene, you expect 1/4 *SS* : 1/2 *Ss* : 1/4 *ss*. To get the final ratios, multiple the frequency of the first genotype by the frequency of the second genotype.

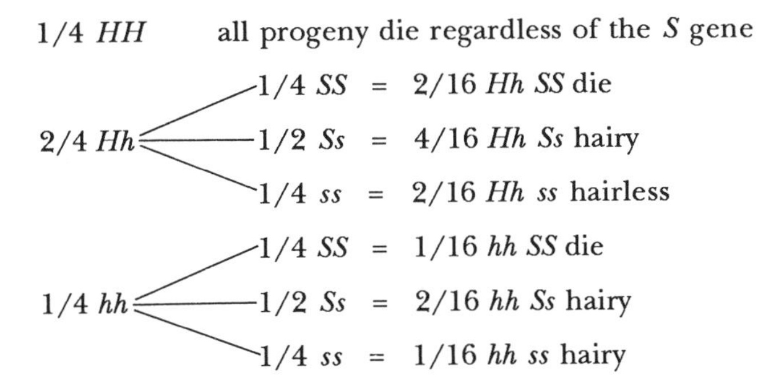

Of the 9 living progeny, the ratio of hairy to hairless is 7:2.

b. This cross is *Hh ss* × *Hh Ss*. A 1:2:1 ratio is expected for the *H* gene and a 1:1 ratio is expected for the *S* gene.

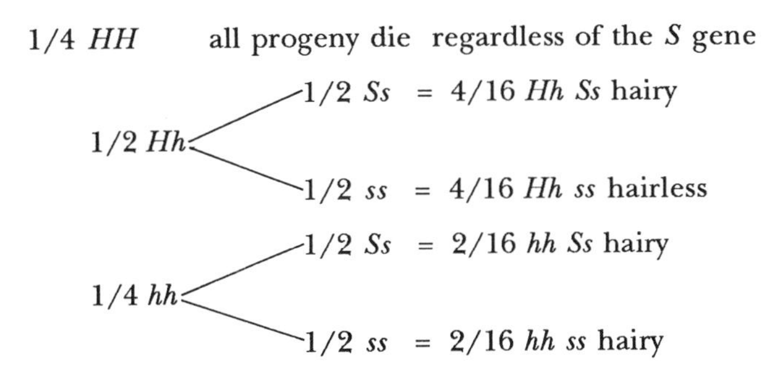

The ratio of hairy to hairless is 2:1.

18. With two independently assorting epistatic genes, you must first state the phenotype associated with each genotype.

B– ee black *B– E–* golden

bb ee brown *bb E–* golden

Cross 1: dog 1 (golden) × dog 2 (golden)

6/8 golden

1/8 black

1/8 brown

The existence of colored dogs (*ee*), black or brown, from two golden parents indicates that the parents were both heterozygous for the *E* gene (*Ee*). The existence of brown indicates that each parent has at least 1 *b* allele. Because black also shows up, at least one parent also has a *B* allele. Because a classical 9:3:3:1 ratio, or some modification of it, is not observed, this is not a dihybrid F_1 cross (not *Bb Ee* × *Bb Ee*). Therefore, the parents are *bb Ee* × *Bb Ee*, but you cannot know yet which parent had which genotype.

Cross 2: dog 1 (golden) × dog 3 (black)

4/8 golden

3/8 black

1/8 brown

Again, color indicates that each parent had at least one *e* allele. The golden parent must be *Ee*, while the black parent is *ee*. This is a testcross with regard to the *E* gene. Thus, 50 percent of the progeny are expected to be golden and 50 percent are expected to be black or brown. The ratio of black to brown is 3:1, indicating a heterozygous (*Bb*) by heterozygous mating. Therefore, dog 1 is *Bb Ee* and dog 3 is *Bb ee*. Dog 2, by a process of elimination from above, is *bb Ee*.

19. Let *A* = white, *a* = yellow, *B* = disk, and *b* = sphere. These symbols were chosen on the basis of the F_1 progeny, which indicate that white and disk are dominant. The F_2 progeny are in the classic 9:3:3:1 ratio, which indicates that they come from a dihybrid cross. Therefore, the crosses conducted were

P *AA BB* × *aa bb*

F_1 *Aa Bb* × *Aa Bb*

F_2 9 *A– B–* : 3 *aa B–* : 3 *A– bb* : 1 *aa bb*

20. Whenever a cross involving two deviants from normal results in a normal phenotype, more than one gene is involved in producing the phenotype, the normal is dominant to the deviation, and the two parents are abnormal for different genes. Thus, one parent could be *aa BB* and the other parent could be *AA bb*. All offspring would be *Aa Bb* (normal). The doubly heterozygous offspring have one copy of a functional allele for each gene, whereas each of the two parents is lacking a functional allele for one of the genes.

21. a. To solve this problem you will have to use a trial and error approach.

The first decision regards the number of genes involved. Cross 2 tells you that there are at least two genes because white X white yields a new phenotype. Cross 5, however, indicates that there are at least three genes involved because a 1:7 (same as 2:14) ratio is not observed with two genes.

Look at cross 1. The two lines cannot compensate for each other, suggesting a shared homozygous defective gene.

Compare crosses 2 and 3. Lines 1 and 3 can compensate for each other's defects but lines 2 and 3 cannot. This suggests that line 2 shares a defective homozygous gene with line 3 and that line 1 is normal for the defective gene seen in line 3 but not in line 2.

At this point, you must arbitrarily make some assumptions and test them against the results. Let the three genes involved be *A/a*, *B/b* and *D/d*. You could assume, for instance, that lines 1 and 2 share a defect in gene *A/a*, lines 2 and 3 share a defect in *B/b*. However, if line 3 can compensate for line 1 but not for line 2, there must be an additional defect in line 2. Thus far, we have determined the genotypes of the lines to be

line 1: *aa BB* ??

line 2: *aa bb dd*

line 3: *AA bb DD*

Because lines 1 and 2 cannot produce color, line 1 must be *aa BB dd*. Now the crosses can be explained.

Cross 1: *aa BB dd* × *aa bb dd* → all *aa Bb dd*

Cross 2: *aa BB dd* × *AA bb DD* → all *Aa Bb Dd*

Cross 3: *aa bb dd* × *AA bb DD* → all *Aa bb Dd*

Cross 4: *Aa Bb Dd* × *aa BB dd* → 1/8 *Aa Bb Dd* red
1/8 *Aa Bb dd* white
1/8 *Aa BB Dd* red
1/8 *Aa BB dd* white
1/8 *aa Bb Dd* white

1/8 *aa Bb dd* white

1/8 *aa BB Dd* white

1/8 *aa BB dd* white

Cross 5: *Aa Bb Dd* × *aa bb dd* → 1/8 *Aa Bb Dd* red

1/8 *Aa Bb dd* white

1/8 *Aa bb Dd* white

1/8 *Aa bb dd* white

1/8 *aa Bb Dd* white

1/8 *aa Bb dd* white

1/8 *aa bb Dd* white

1/8 *aa bb dd* white

Cross 6: *Aa Bb Dd* × *AA bb DD* → 1/8 *AA Bb Dd* red

1/8 *Aa Bb Dd* red

1/8 *AA Bb DD* red

1/8 *Aa Bb DD* red

1/8 *AA bb DD* white

1/8 *Aa bb DD* white

1/8 *AA bb Dd* white

1/8 *Aa bb Dd* white

b. The cross is *Aa Bb Dd* × *Aa bb Dd*. The red progeny will have to be *A– B– D–*, which equals (3/4)(1/2)(3/4) = 9/32.

22. The first step in each cross is to write as much of the genotype as possible from the phenotype.

Cross 1: *A– B–* × *aa bb* → 1 *A– B–* : 2 ?– ?– : 1 *aa bb*

Because the double recessive appears, the blue parent must be *Aa Bb*. The two purple then must be *Aa bb* and *aa Bb*.

Cross 2: ?? ?? × ?? ?? → 1 *A– B–* : 2 ?? ?? : 1 *aa bb*

The two parents must be, in either order, *Aa bb* and *aa Bb*. The two purple progeny must be the same. The blue progeny are *Aa Bb*.

Cross 3: *A– B–* × *A– B–* → 3 *A– B–* : 1 ?? ??

The only conclusions possible here are that one parent is either *AA* or *BB* and the other parent is *Bb* if the first is *AA* or *Aa* if the first is *BB*.

Cross 4: $A\!-\ B\!-\times$?? ?? $\rightarrow$ 3 $A\!-\ B\!-$: 4 ?? ?? : 1 *aa bb*

The purple parent can be either *Aa bb* or *aa Bb* for this answer. Assume the purple parent is *Aa bb*. The blue parent must be *Aa Bb*. The progeny are

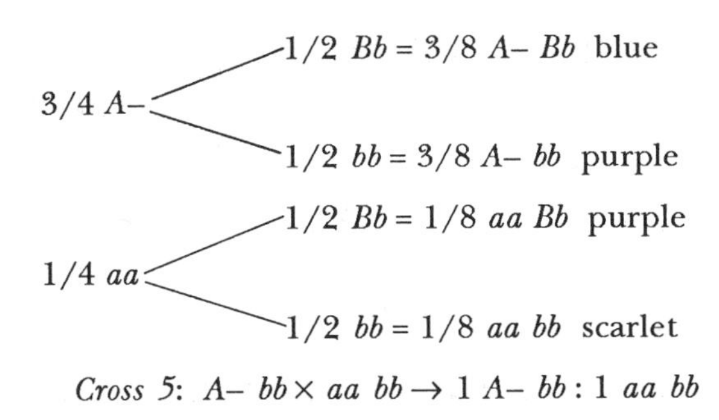

Cross 5: $A\!-\ bb \times aa\ bb \rightarrow 1\ A\!-\ bb : 1\ aa\ bb$

As written this is, in essence, a testcross for gene *A/a*. The purple parent and progeny are *Aa bb*. Alternatively, the purple parent and progeny could be *aa Bb*.

23. The F_1 progeny of cross 1 indicate that sun-red is dominant to pink. The F_2 progeny, which are approximately in a 3:1 ratio, support this. The same pattern is seen in crosses 2 and 3, with sun-red dominant to orange and orange dominant to pink. Thus, we have a multiple allelic series with sun-red > orange > pink. In all three crosses, the parents must be homozygous.

If c^{sr} = sun-red, c^{O} = orange, and c^{p} = pink, then the crosses and the results are

Cross 1: $c^{sr}c^{sr} \times c^{p}c^{p}$ gives F_1 $c^{sr}c^{p}$; F_2 $3\ c^{sr}\!- : 1\ c^{p}c^{p}$

Cross 2: $c^{O}c^{O} \times c^{sr}c^{sr}$ gives F_1 $c^{sr}c^{O}$; F_2 $3\ c^{sr}\!- : 1\ c^{O}c^{O}$

Cross 3: $c^{O}c^{O} \times c^{p}c^{p}$ gives F_1 $c^{O}c^{p}$; F_2 $3\ c^{O}\!- : 1\ c^{p}c^{p}$

Cross 4 presents a new situation. The color of the F_1 differs from that of either parent, suggesting that two separate genes are involved. An alternative explanation is either codominance or incomplete dominance. If either codominance or incomplete dominance is involved, then the F_2 will appear in a 1:2:1 ratio. If two genes are involved, then a 9:3:3:1 ratio, or some variant of it, will be observed. The progeny actually are in a 9:4:3 ratio. This means that two genes are involved and that there is epistasis.

Furthermore, the two F_1 parents must have been heterozygous for three phenotypes to be present in the F_2.

Let a stand for the scarlet gene and A for its colorless allele, and assume that there is a dominant allele, C, that blocks the expression of the gene that we have been studying to this point.

Cross 4: P $c^O c^O\ AA \times CC\ aa$

F_1 $Cc^O\ Aa$

F_2 9 $C\text{–}\ A\text{–}$ yellow

3 $C\text{–}\ aa$ scarlet

3 $c^O c^O A\text{–}$ orange

1 $c^O c^O\ aa$ orange (epistasis, with c^O blocking the expression of aa).

24. a. P AA (agouti) × aa (nonagouti)

gametes A and a

F_1 Aa (agouti)

gametes A and a

F_2 1 AA (agouti) : 2 Aa (agouti) : 1 aa (nonagouti)

b. P BB (wild-type) × bb (cinnamon)

gametes B and b

F_1 Bb (wild-type)

gametes B and b

F_2 1 BB (wild-type) : 2 Bb (wild-type) : 1 bb (cinnamon)

c. P $AA\ bb$ (cinnamon or brown agouti) × $aa\ BB$ (black nonagouti)

gametes Ab and aB

F_1 $Aa\ Bb$ (wild type or black agouti)

d. 9 $A\text{–}\ B\text{–}$ black agouti

3 $aa\ B\text{–}$ black nonagouti

3 $A\text{–}\ bb$ cinnamon

1 $aa\ bb$ chocolate

e. P *AA bb* (cinnamon) × *aa BB* (black nonagouti)

gametes *Ab* and *aB*

F_1 *Aa Bb* (wild type)

gametes *AB*, *Ab*, *aB*, and *ab*

F_2 9 *A– B–* wild type

1 *AA BB*

2 *Aa BB*

2 *AA Bb*

4 *Aa Bb*

3 *aa B–* black nonagouti

1 *aa BB*

2 *aa Bb*

3 *A– bb* cinnamon

1 *AA bb*

2 *Aa bb*

1 *aa bb* chocolate

f. P *Aa Bb* × *AA bb* (wild-type) (cinnamon)

F_1 1 *AA Bb* wild-type

1 *Aa Bb* wild-type

1 *AA bb* cinnamon

1 *Aa bb* cinnamon

Aa Bb × *aa BB* (wild-type) (black nonagouti)

1 *Aa BB* wild-type

1 *Aa Bb* wild-type

1 *aa BB* black nonagouti

1 *aa Bb* black nonagouti

g. P *Aa Bb* × *aa bb* (wild-type) (chocolate)

F_1 1 *Aa Bb* wild-type

1 *Aa bb* cinnamon

1 *aa Bb* black nonagouti

1 *aa bb* chocolate

h. To be albino, the mice must be *cc*, but the genotype with regard to the *A/a* and *B/b* genes can be determined only by realizing that the wild-type is *AA BB* and looking at the F_2 progeny.

Cross 1:	P	*cc* ?? ?? × *CC AA BB*
	F_1	*Cc A– B–*
	F_2	87 wild-type *C– A– B–*
		32 cinnamon *C– A– bb*
		39 albino *cc* ?? ??

For cinnamon to appear in the F_2, the F_1 parents must be *Bb*. Because the wild-type is *BB*, the albino parent must have been *bb*. Now the F_1 parent can be written *Cc A– Bb*. With such a cross, one-fourth of the progeny would be expected to be albino (*cc*), which is what is observed. Three-fourths of the remaining progeny would be black, either agouti or nonagouti, and one-fourth would be either cinnamon, if agouti, or chocolate, if nonagouti. Because chocolate is not observed, the F_1 parent must not carry the allele for nonagouti. Therefore, the F_1 parent is *AA* and the original albino must have been *cc AA bb.*

Cross 2:	P	*cc* ?? ?? X *CC AA BB*
	F_1	*Cc A– B–*
	F_2	62 wild-type *C– A– B–*
		18 albino *cc* ?? ??

This is a 3:1 ratio, indicating that only one gene is heterozygous in the F_1. That gene must be *Cc*. Therefore, the albino parent must be *cc AA BB*.

Cross 3:	P	*cc* ?? ?? × *CC AA BB*
	F_1	*Cc A– B–*
	F_2	96 wild-type *C– A– B–*
		30 black *C– aa B–*
		41 albino *cc* ?? ??

For a black nonagouti phenotype to appear in the F_2, the F_1 must have been heterozygous for the *A/a* gene. Therefore, its genotype can be written *Cc Aa B–* and the albino parent must be *cc aa* ??. Among the colored F_2 a 3:1 ratio is observed, indicating that only one of the two genes is heterozygous in the F_1. Therefore, the F_1 must be *Cc Aa BB* and the albino parent must be *cc aa BB*.

Cross 4:	P	*cc* ?? ?? × *CC AA BB*
	F_1	*Cc A– B–*
	F_2	287 wild-type *C– A– B–*
		86 black *C– aa B–*
		92 cinnamon *C– A– bb*
		29 chocolate *C– aa bb*
		164 albino *C–* ?? ??

To get chocolate F_2 progeny the F_1 parent must be heterozygous for all genes and the albino parent must be *cc aa bb*.

25. To solve this problem, first restate the information:

A– yellow	*A– R–* gray
R– black	*aa rr* white

The cross is gray × yellow, or *A– R–* × *A– rr*. The F_1 progeny are

3/8 yellow	1/8 black
3/8 gray	1/8 white

To achieve white, both parents must carry an *r* and an *a* allele. Now the cross can be rewritten as *Aa Rr* × *Aa rr*.

26. Whenever two lines that are both deviant from wild-type produce wild-type there must be two genes involved in the characteristic and each line is abnormal for one of them. Furthermore, the deviation is recessive in both cases. Let scarlet be represented by *aa BB* and dark brown be represented by *AA bb*. The F_1 progeny must be *Aa Bb*. The F_2 progeny are

432 red	*A– B–*	139 brown	*AA bb*
158 scarlet	*aa B–*	52 white	?? ??

Notice that this is a 9:3:3:1 ratio, indicating a dihybrid cross. The white phenotype must therefore be *aa bb*.

27. a. P single-combed × walnut-combed

(*rr pp*) (*RR PP*)

F_1 *Rr Pp* walnut

F_2 9 *R– P–* walnut

3 *rr P–* pea

3 *R– pp* rose

1 *rr pp* single

b. P walnut-combed × rose-combed

(*R– P–*) (*R– pp*)

F_1 3/8 *R– pp* rose

3/8 *R– P–* walnut

1/8 *rr P–* pea

1/8 *rr pp* single

The 3:1 *R–* : *rr* ratio indicates that the parents were heterozygous. The 1:1 *P–* : *pp* ratio indicates a testcross for this gene, *Pp* × *pp*. Therefore, the parents were *Rr Pp* and *Rr pp*.

c. P walnut-combed × rose-combed

(*R– P–*) (*R– pp*)

F_1 walnut

(*R– P–*)

To get this result, one of the parents must be homozygous *R*, but both need not be, and the walnut parent must be homozygous *PP*.

d. *RR PP, Rr PP, RR Pp, Rr Pp*

28. A trihybrid F_1 cross produces a 27:9:9:9:3:3:3:1 ratio, which could look like a 27:37 ratio with epistasis. The 27 would come from all genotypes that are *A– B– D–*, while any genotype homozygous for one or more recessive alleles would have the alternative phenotype.

29. a. Two deviations from normal that give wild-type when crossed are indicative of two genes and recessive deviations. Because the males are wild-type, the defect in line B is autosomal. Therefore, the cross can be represented at this point as

P line-A males × line-B females

d– EE × *DD ee*

F_1 200 wild-type males *D– Ee*

198 wild-type females *Dd Ee*

where *d* is the defective allele in line A and *e* is the defective allele in line B. The dash indicates a lack of knowledge of the location of the *D/d* gene (autosomal or X-linked).

b. The reciprocal cross indicates that the defect in line A is X-linked.

c.

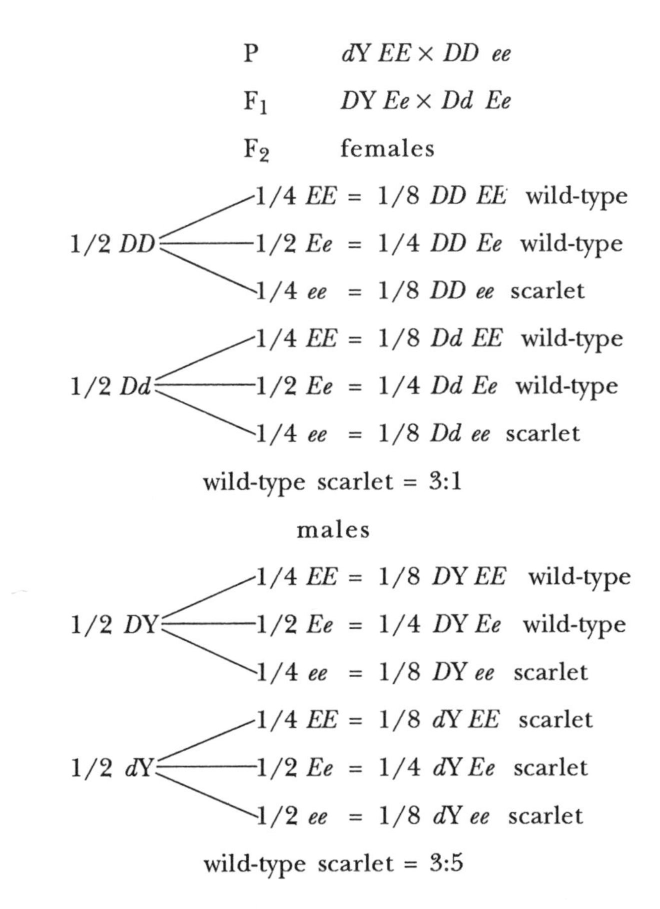

P *dY EE* × *DD ee*

F_1 *DY Ee* × *Dd Ee*

F_2 females

1/2 *DD* — 1/4 *EE* = 1/8 *DD EE* wild-type
— 1/2 *Ee* = 1/4 *DD Ee* wild-type
— 1/4 *ee* = 1/8 *DD ee* scarlet

1/2 *Dd* — 1/4 *EE* = 1/8 *Dd EE* wild-type
— 1/2 *Ee* = 1/4 *Dd Ee* wild-type
— 1/4 *ee* = 1/8 *Dd ee* scarlet

wild-type scarlet = 3:1

males

1/2 *DY* — 1/4 *EE* = 1/8 *DY EE* wild-type
— 1/2 *Ee* = 1/4 *DY Ee* wild-type
— 1/4 *ee* = 1/8 *DY ee* scarlet

1/2 *dY* — 1/4 *EE* = 1/8 *dY EE* scarlet
— 1/2 *Ee* = 1/4 *dY Ee* scarlet
— 1/2 *ee* = 1/8 *dY ee* scarlet

wild-type scarlet = 3:5

30. a. For the same reasons stated in Problem 23a, two genes are involved, both deviations are recessive, and the line-B defect is autosomal. The cross can be represented as

P line-A males × line-B females

d– EE × *DD ee*

F_1 435 wild-type males *D– Ee*

428 wild-type females *Dd Ee*

where *d* is the defective allele in line A, *e* is the defective allele in line B and the dash indicates a lack of information with regard to the location of the line-A defect.

b. The appearance of white-eyed males in the F_1 indicates that the line A defect is X-linked. The cross can be represented as

P line-B males × line-A females

DY ee × *dd EE*

F_1 420 white-eyed males *dY Ee*

405 wild-type females *Dd Ee*

c. The F_1 cross in part **a** should be *D* Y *Ee* × *Dd Ee*, yielding a 3 wild-type : 5 scarlet ratio in males and a 3 wild-type : 1 scarlet ratio in females. Because the predictions do not match the results, the cross as indicated must be wrong.

Among females a 9:3:3:1 result was obtained. This indicates that a third gene is involved with eye color. It must be autosomal because the males are also in a 9:3:3:1 ratio. Therefore, the cross in part a should be rewritten as

P *dY EE FF* × *DD ee ff* (or *dY EE ff* × *DD ee FF*)

F_1 *DY Ee Ff* × *Dd Ee Ff*.

F_2 females

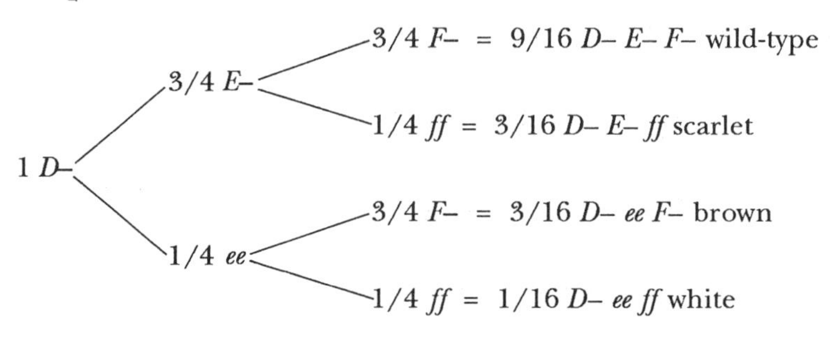

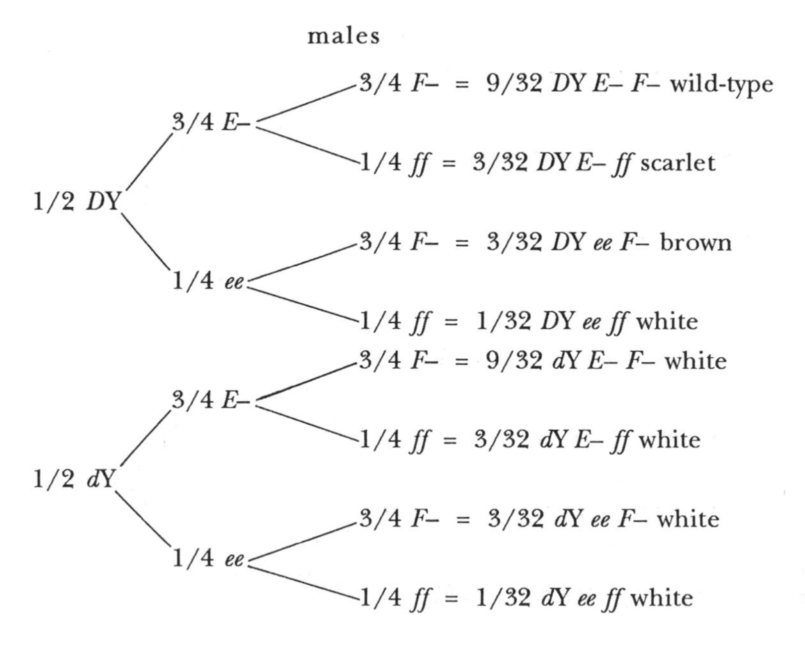

The cross in part **b** is

P *DY ee ff* × *dd EE FF* (or *DY ee FF* × *dd EE ff*)

F_1 *dY Ee Ff* × *Dd Ee Ff*

F_2 females

1/2 *D–*
- 3/4 *E–*
  - 3/4 *F–* = 9/32 *D– E– F–* wild-type
  - 1/4 *ff* = 3/32 *D– E– ff* scarlet
- 1/4 *ee*
  - 3/4 *F–* = 3/32 *D– ee F–* brown
  - 1/4 *ff* = 1/32 *D– ee ff* white

1/2 *dd*
- 3/4 *E–*
  - 3/4 *F–* = 9/32 *dd E– F–* white
  - 1/4 *ff* = 3/32 *dd E– ff* white
- 1/4 *ee*
  - 3/4 *F–* = 3/32 *dd ee F–* white
  - 1/4 *ff* = 1/32 *dd ee ff* white

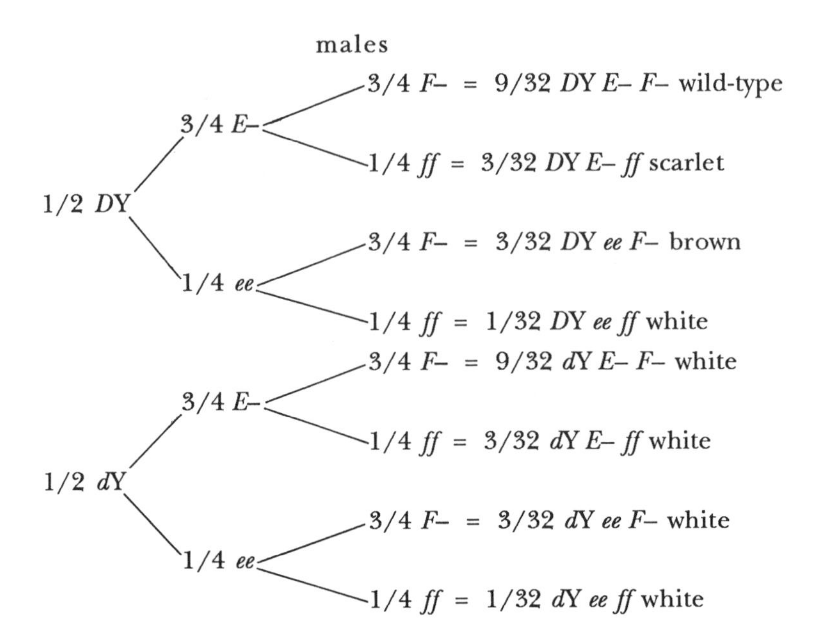

31. The results indicate that two genes are involved (modified 9:3:3:1 ratio), with white blocking the expression of color by the other gene. The ratio of white : color is 3:1, indicating that the F_1 is heterozygous (*Ww*). Among colored dogs, the ratio is 3 black : 1 brown, indicating that black is dominant to brown and the F_1 is heterozygous (*Bb*). The original brown dog is *ww bb* and the original white dog is *WW BB.* The F_1 progeny are *Ww Bb* and the F_2 progeny are

9 *W– B–*	white	3 *ww B–*	black
3 *W– bb*	white	1 *ww bb*	brown

32.

Cross	Results	Conclusion
A- C- R- × *aa cc RR*	50% colored	Colored or white will depend on the *A* and *C* genes. Because half the seeds are colored, one of the two genes is heterozygous.
A- C- R- × *aa CC rr*	25% colored	Color depends on *A* and *R* here. If only one gene were heterozygous, 50% would be colored. Therefore, both *A* and *R*

		are heterozygous. The seed is *Aa CC Rr*.
A- C- R- × *AA cc rr*	50% colored	This supports the above conclusion.

33. a. The cross is

P *td su* (wild-type) × td^+ su^+ (wild-type)

F_1 1 *td su* wild-type

1 *td* su^+ requires tryptophan

1 td^+ su^+ wild-type

1 td^+ *su* wild-type

b. 1 tryptophan-dependent : 3 tryptophan-independent

34. *Cross 1*:

P *Ww Bb Oo* (white) × *Ww Bb Oo* (white)

F_1 3/4 *W–* ?? ?? white (48/64)

1/4 *WW* ?? ?? lethal (16/64)

1/2 *Ww* ?? ?? white (32/64)

1/4 *ww* ?? ?? colored (16/64)

Among the colored animals, the following progeny will be found:

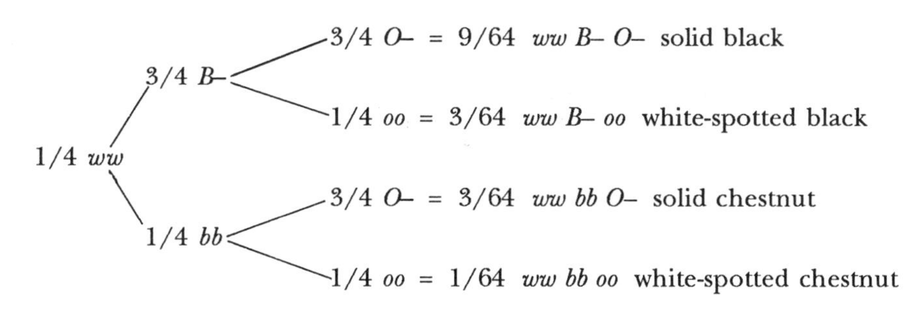

Cross 2:

P *Ww Bb Oo* (white) × *Ww Bb oo* (white)

F_1 3/4 *W–* ?? ?? white (24/32)

1/4 *WW* ?? ?? lethal (8/32)

1/2 *Ww* ?? ?? white (16/32)

1/4 *ww* ?? ?? colored (8/32)

Among the colored animals, the following progeny will be found:

3/32	*ww B– O–*	solid black
3/32	*ww B– oo*	white-spotted black
1/32	*ww bb O–*	solid chestnut
1/32	*ww bb oo*	white-spotted chestnut

35. P *AA BB CC DD SS* × *aa bb cc dd ss*

F_1 *Aa Bb Cc Dd Ss*

F_2

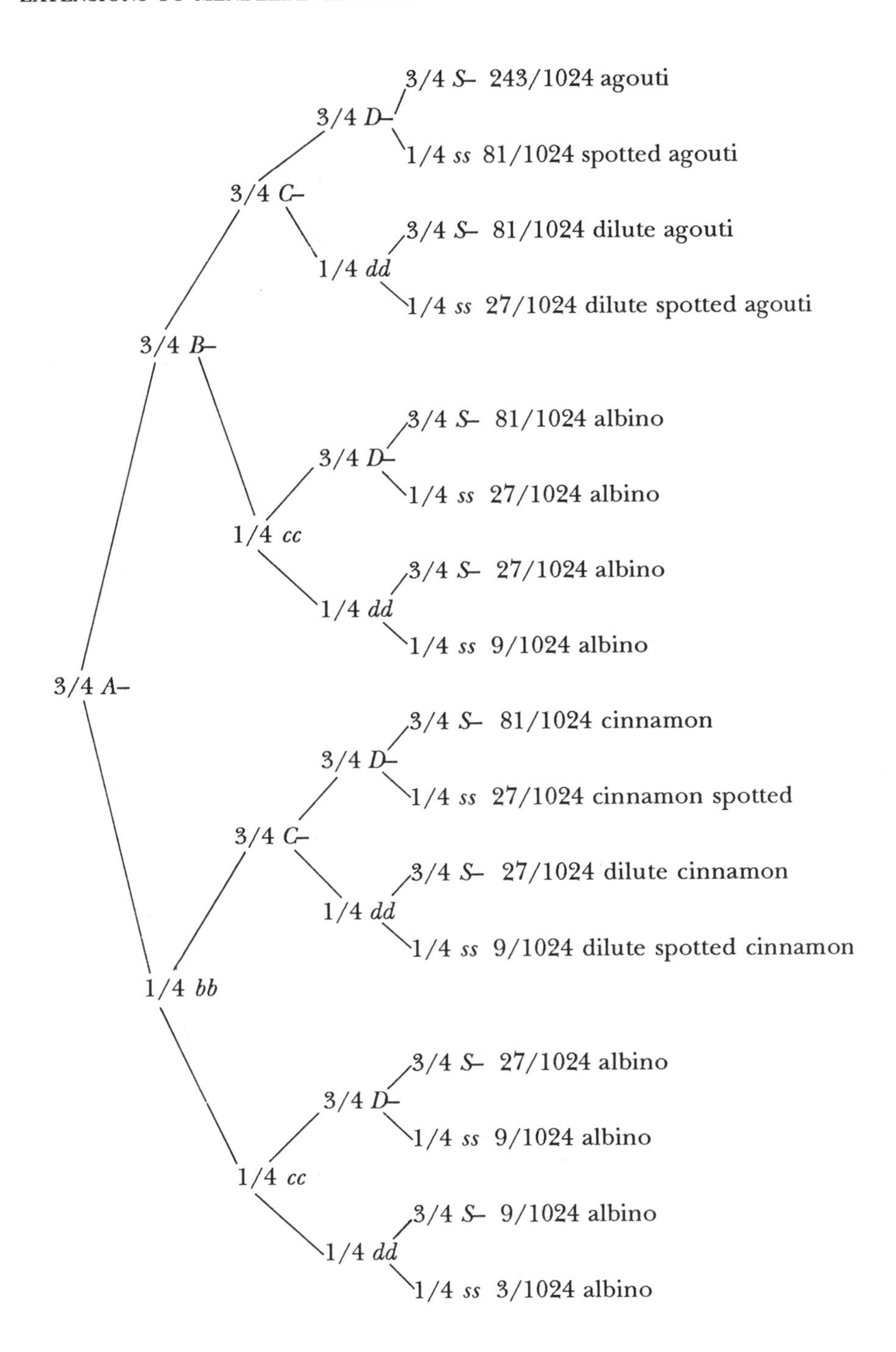
3/4 A–
3/4 B–
3/4 C–
3/4 D–
3/4 S– 243/1024 agouti
1/4 ss 81/1024 spotted agouti
1/4 dd
3/4 S– 81/1024 dilute agouti
1/4 ss 27/1024 dilute spotted agouti
1/4 cc
3/4 D–
3/4 S– 81/1024 albino
1/4 ss 27/1024 albino
1/4 dd
3/4 S– 27/1024 albino
1/4 ss 9/1024 albino
1/4 bb
3/4 C–
3/4 D–
3/4 S– 81/1024 cinnamon
1/4 ss 27/1024 cinnamon spotted
1/4 dd
3/4 S– 27/1024 dilute cinnamon
1/4 ss 9/1024 dilute spotted cinnamon
1/4 cc
3/4 D–
3/4 S– 27/1024 albino
1/4 ss 9/1024 albino
1/4 dd
3/4 S– 9/1024 albino
1/4 ss 3/1024 albino

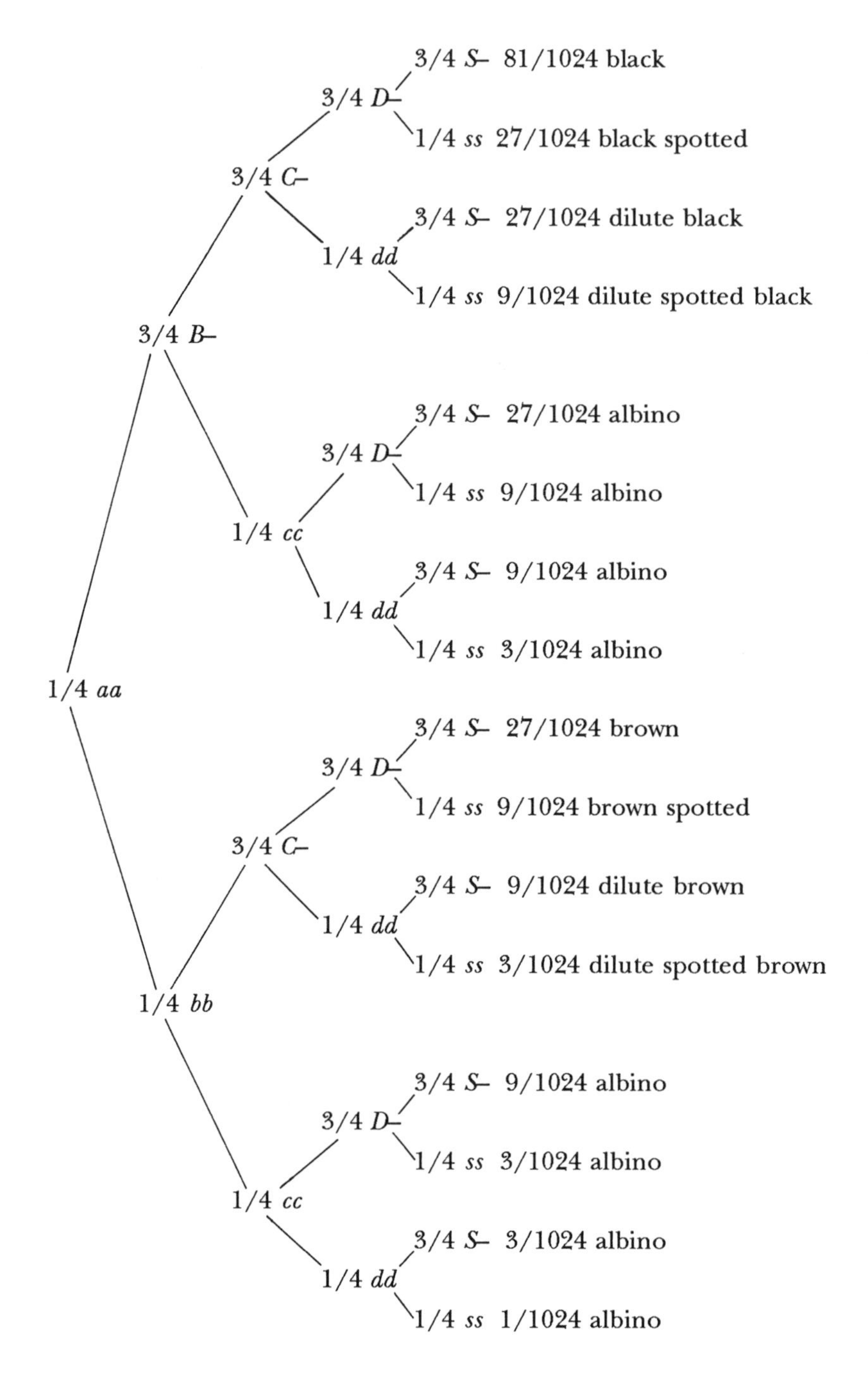

36. Both the F_1 resistance to the two fungus diseases and the F_2 ratio indicate that two genes are involved in disease resistance, that resistance is dominant to susceptibility, and that the genes are autosomal.

Let D = resistance to α and E = resistance to ß.

a.

P	*dd EE* × *DD ee*
F_1	*Dd Ee*
F_2	9 *D– E–* unaffected
	3 *dd E–* diseased
	3 *D– ee* diseased
	1 *dd ee* diseased

b. If virulence is due to two separate genes

let F = virulent in variety A

f = nonvirulent

G = virulent in variety B

g = nonvirulent

Then the cross is *Fg* (α) × *fG* (ß). Progeny would be of four types:

Fg = disease in A

fg = nonvirulent in A and B

FG = virulent in A and B

fG = virulent in B

In a test against the two varieties, the following results would be obtained, where + indicates infection:

	A	B
Fg	+	–
fg	–	–
FG	+	+
fG	–	+

If virulence is due to one gene, with two alleles,

let F = virulent in A

F' = virulent in B

Progeny from a cross would be F or F' in a 1:1 ratio.

In a test against the two varieties, the following results would be obtained, where + indicates infection:

	A	B
F	+	–
F′	–	+

37. Pedigrees like this are quite common. They indicate lack of penetrance due to epistasis or environmental effects.

38. If **a**: To demonstrate allelic alternatives, typical Mendelian crosses need to be done. Attempt to prove that each plant type is true-breeding by selfing. Then cross each plant type with each of the others, proceeding at least through the F_2. The observed ratios will allow you to make a decision.

If **b**: Crosses like the above will indicate, through the observed ratios, whether other genes are modifying the expression of spots.

If **c**: If the variable expressivity is due to the environment, then cuttings grown from the three plants in controlled environments will all have the same phenotype.

39. In cross 1, the following can be written immediately:

P *M– D– ww* (dark magenta) × *mm* ?? ?? (white with yellowish spots)

F_1 1/2 *M– D– ww* dark magenta

1/2 *M– dd ww* light magenta

All progeny are colored, indicating that no *W* allele is present in the parents. Because the progeny are in a 1:1 ratio, only one of the genes in the parents is heterozygous. Also, the light magenta progeny, *dd*, indicates which gene that is. Therefore, the genotypes must be

P *MM Dd ww* X *mm dd ww*

F_1 1 *Mm Dd ww* : 1 *Mm dd ww*

In cross 2, the following can be written immediately:

P *mm* ?? ?? (white with yellowish spots) × *M– dd ww* (light magenta)

F_1 1/2 *M–* ?? *W–* white with magenta spots

1/4 *M– D– ww* dark magenta

1/4 *M– dd ww* light magenta

For light and dark plants to appear in a 1:1 ratio among the colored plants, one of the parents must be heterozygous *Dd*. The ratio of white to colored is 1:1, a testcross, so one of the parents is heterozygous *Ww*. All plants are magenta, indicating that one parent is homozygous *MM*. Therefore, the genotypes are

P *mm Dd Ww* × *MM dd ww*

F_1 1/2 *Mm Dd* (or *dd*) *Ww* (or *ww*)

1/4 *Mm Dd ww*

1/4 *Mm dd ww*

Note that two genotypes were determined for white plants with yellowish spots in the two crosses: *mm dd ww* and *mm Dd ww*. Both are correct. Because the plants do not make pigment, the enhancer (*D–* or *dd*) gene is irrelevant to the phenotype.

40. a. This is a dihybrid cross with only one phenotype in the F_2 being colored. The ratio of white to red indicates that the double recessive is not the colored phenotype. Instead, the general formula for color is represented by *X– yy*.

Let line 1 be *AA BB* and line 2 be *aa bb*. The F_1 is *Aa Bb*. Assume that *A* blocks color in line 1 and *bb* blocks color in line 2. The F_1 will be white because of the presence of *A*. The F_2 are

9 *A– B–* white because of *A*

3 *A– bb* white because of *A*

3 *aa B–* red

1 *aa bb* white because of *bb*

b. *Cross 1*: *AA BB* × *Aa Bb* → all *A– B–* white

Cross 2: *aa bb* × *Aa Bb* → 1/4 *Aa Bb* white

1/4 *Aa bb* white

1/4 *aa bb* white

1/4 *aa Bb* red

41. Remember that monozygotic twins are genotypically identical. Therefore, any differences between them must be due to developmental noise. Also, left-right differences within an individual are considered to be due to developmental noise. This family shows some indications of a genetic component to fingerprint patterns, but no conclusions can be drawn from a sample consisting of only one family.

42. a. Note that blue is always present, indicating *EE* (blue) in both parents. Because of the ratios that are observed neither C nor D are varying. In this case, the gene pairs that are involved are A/a and B/b. The F_1 is *Aa Bb* and the F_2 is

9 *A– B–*	blue + red, or purple
3 *A– bb*	blue + yellow, or green
3 *aa B–*	blue + $white_2$, or blue
1 *aa bb*	blue + $white_2$, or blue

b. Blue is not always present, indicating *Ee* in the F_1. Because green never appears, the F_1 must be *BB CC DD.* The F_1 is *Aa Ee,* and the F_2 is

9 *A– E–*	red + blue, or purple
3 *A– ee*	red + $white_1$, or red
3 *aa E–*	$white_2$ + blue, or blue
1 *aa ee*	$white_2$ + $white_1$, or white

c. Blue is always present, indicating that the F_1 is *EE.* No green appears, indicating that the F_1 is *BB.* The two genes involved are *A/a* and *D/d.* The F_2 is

9 *A– D–*	blue + red + $white_4$, or purple
3 *A– dd*	blue + red, or purple
3 *aa D–*	blue + $white_2$ + $white_4$, or blue
1 *aa dd*	$white_2$ + blue + red, or purple

d. The presence of yellow indicates *bb ee* in the F_2. Therefore, the F_1 is *Bb Ee* and the F_2 is

9 *B– E–*	red + blue, or purple
3 *B– ee*	red + $white_1$, or red
3 *bb E–*	yellow + blue, or green
1 *bb ee*	yellow + $white_1$, or yellow

Tips on Problem Solving

Whenever a cross involving two deviants from normal results in a normal phenotype, more than one gene is involved in producing the phenotype, the normal is dominant to the deviation, and the two parents are abnormal for different genes (see Problem 20).

Whenever the sexes differ with respect to phenotype among the progeny, an X-linked gene is involved (see Problem 16).

If the F_1 phenotype differs from that of either parent, two separate genes may be involved. An alternative explanation is either codominance or incomplete dominance. If either codominance or incomplete dominance is involved, then the F_2 progeny will appear in a 1:2:1 ratio. If two genes are involved, then a 9:3:3:1 ratio, or some variant of it, will be observed (see Problem 23).

Self-Test

1. Discuss the difference between dominance and epistasis.

2. What are the factors that lead to a lack of penetrance?

3. What are the factors that lead to variable expressivity?

4. How can you distinguish between multiple alleles and multiple genes affecting a characteristic?

5. What is the maximum number of phenotypes that can be observed when the gene in question is multiply allelic?

6. In a pedigree, how can you distinguish between recessiveness and a lack of penetrance of a dominant disorder?

7. You have two pure-breeding lines of *Drosophila*, lines A and B, which have scarlet eyes. Answer the following questions in sequence.

a. A cross of line-A females with line-B males yields all wild-type offspring. What can you conclude?

b. The reciprocal cross yields wild-type females and scarlet males. What can you conclude?

c. When you cross F_1 females from part **a** with F_1 males from part **b**, what phenotypic ratio is observed?

8. In species X, both *AA* and *aa* individuals die. Another independent gene, *B*/*b*, blocks the lethality of only the *AA* genotype. Otherwise, it has no effect. What genotypic ratio among the viable progeny would be observed in a cross of *Aa Bb* with *Aa BB*?

9. What genotypic ratio is observed in a testcross of a triple hybrid?

10. Consider the dihybrid cross *Aa Bb* × *Aa Bb*. Among the 16-offspring matings, what is the percentage that would be expected to have exactly 9 *A– B–*?

11. Discuss the following pedigree, where the symbols indicate:

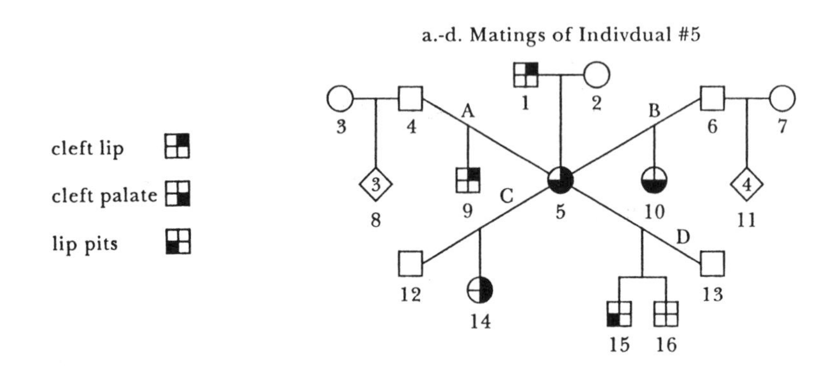

Solutions to Self-Test

1. Dominance refers to relationships between alleles, and epistasis refers to relationships between genes.

2. A lack of penetrance can be due either to the environment or to the expression of other genes.

3. Variable expressivity can be due either to the environment or to the expression of other genes.

4. Multiple alleles show a 1:2:1 ratio or some modification of that ratio. Multiple genes show a 9:3:3:1 ratio or some modification of that ratio.

5. For any specific cross, the maximum number of phenotypes is four.

6. It is often quite difficult to distinguish between recessiveness and a lack of penetrance of a dominant disorder. In practice, those disorders that have low penetrance are often ascribed to the interaction of multiple genes (epistasis), and risk values are based upon empiric observations

rather than upon Mendelian ratios. The same is true of recessive disorders.

7. a. Two genes govern eye color. The defects are both recessive. The defect in line A is autosomal.

b. The defect in line B is X-linked. The two crosses can be represented as

Cross 1: P *dd EE* × *DD eY*

F_1 *Dd Ee* females

Dd EY males

Cross 2: P *dd EY* × *DD ee*

F_1 *Dd Ee* females

Dd eY males

c. The cross is

P *Dd Ee* × *Dd eY*

F_1 females

3/4 *D–* → 1/2 *Ee* = 3/8 *D– E–* wild-type

3/4 *D–* → 1/2 *ee* = 3/8 *D– ee* scarlet

1/4 *dd* → 1/2 *Ee* = 1/8 *dd E–* scarlet

1/4 *dd* → 1/2 *ee* = 1/8 *dd ee scarlet*

3 wild-type : 5 scarlet

males

3/4 *D–* → 1/2 *EY* = 3/8 *D– EY* wild-type

3/4 *D–* → 1/2 *eY* = 3/8 *D– eY* scarlet

1/4 *dd* → 1/2 *EY* = 1/8 *dd EY* scarlet

1/4 *dd* → 1/2 *eY* = 1/8 *dd eY* scarlet

3 wild-type : 8 scarlet

8. P *Aa Bb* × *Aa BB*

F_1

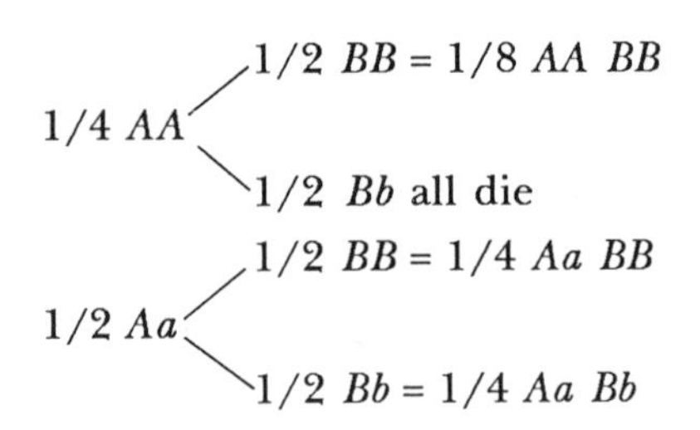

1/4 *aa* all die

Because only 5/8 of the progeny survive, the genotypic ratio of the viable progeny is

1/5 *AA BB*

2/5 *Aa BB*

2/5 *Aa Bb*

9. The cross is *Aa Bb Dd* × *aa bb dd.* The gametes from the second parent *a b d*, and from the first parent are

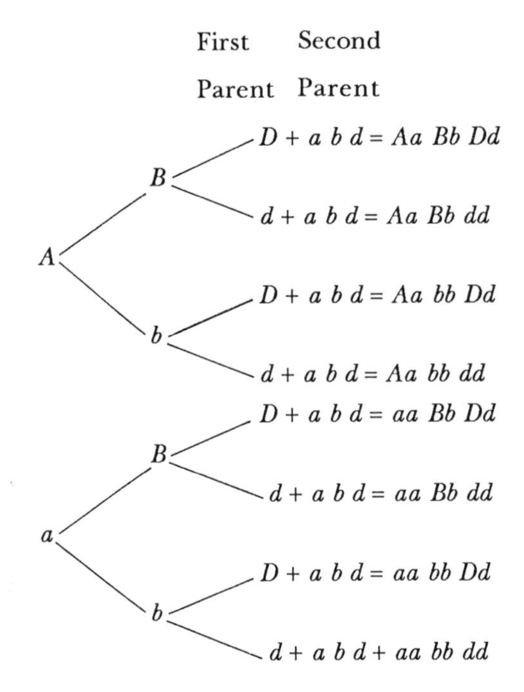

All gametes occur at a frequency of 1/8.

10. The formula to use is

$$\frac{n!}{p!\,q!}(r)^{p}(s)^{q}$$

where n is the total number, p is the number of one kind, q is the number of the alternative, r is the probability of p occurring and s is the probability of q occurring. The exclamation point indicates factorial. $5! = 5 \times 4 \times 3 \times 2 \times 1$.

The probability of A– B– is $(3/4)(3/4) = 9/16$. The probability of all other outcomes is $1 - (3/4)(3/4) = 7/16$. The answer is

$$16!(9/16)^{9}(7/16)^{7}/(9!)(7!).$$

11. Disturbance in the formation of the lip and palate is dominant and probably autosomal. A pleiotropic gene leads to the specific symptoms of the cleft lip and palate syndrome, showing reduced penetrance and variable expressivity.

A Systematic Approach to Problem Solving

Now that you have struggled with a number of genetics problems, it may be worthwhile to make some generalizations about problem solving beyond what has been presented for each chapter so far.

The first task always is to determine exactly what information has been presented and what is being asked. Frequently, it is necessary to rewrite the problem or to symbolize the presented information in some way.

The second task is to formulate and test hypotheses. If the results generated by a hypothesis contradict some aspect of the problem, then the hypothesis is rejected. If the hypothesis generates data compatible with the problem, then it is retained.

A systematic approach is the only safe approach in working genetics problems. Shortcuts in thought processes usually lead to an incorrect answer.

Consider the following two types of problems

1. When analyzing pedigrees, there are only four possibilities (hypotheses) to be considered: autosomal dominant, autosomal recessive, X-linked dominant, and X-linked recessive. The criteria for each should be checked against the data. Additional factors that should be kept in mind are: epistasis, penetrance, expressivity, age of onset, incorrect diagnosis in earlier generations, adultery, adoptions that are not mentioned, and inaccurate information in general. All of these factors can be expected in real life, although few will be encountered in the problems presented here.

2. When studying matings, frequently the first task is to decide whether you are dealing with one gene, two genes, or more than two genes (hypotheses). The location of the gene(s) may or may not be important. If location is important, then there are two hypotheses: autosomal and X-linked. If there are two or more genes, then you may have to decide on linkage relationships between them. There are two hypotheses: unlinked and linked.

If ratios are presented, then 1:2:1 (or some modification signaling dominance) indicates one gene, 9:3:3:1 (or some modification reflecting epistasis) indicates two genes, and 27:9:9:9:3:3:1 (or some modification signaling epistasis) indicates three genes. If ratios are presented that bear no relationship to the above, such as 35:35:15:15, then you are dealing with two linked genes.

If phenotypes rather than ratios are emphasized in the problem, then a cross of two mutants that results in wild-type indicates two genes rather than alleles of the same gene. Both mutants are recessive to wild-type. A correlation of sex with phenotype indicates X-linkage of the gene mutant in the female parent, while a lack of correlation indicates autosomal location.

If the problem involves X-linkage, frequently the only way to solve it is to focus on the male progeny.

Once you determine the number of genes being followed and their location, the problem essentially solves itself if you make a systematic listing of genotype and phenotype.

Sometimes, the final portion of the problem will give additional information that requires you to adjust all the work that you have done up to that point. As an example, in Problem 4–30, parts a and b led you to assume that you were working with two genes. In part c, data incompatible with this assumption were presented. Your initial assumption of two genes was correct for the information given in the first two parts; it was not a mistake.

Other than a lack of systematic thought, the greatest mistake that a student can make is to label a rejected hypothesis an error. This decreases self-confidence and increases anxiety, with the result that real mistakes will likely follow. The beginner needs to keep in mind that science progresses by the rejection of hypotheses. When a hypothesis is rejected, something concrete is known: the proposed hypothesis does not explain the results. An unrejected hypothesis may be right or it may be wrong, and there is no way to know without further experimentation.

A very generalized flowchart for problem-solving would look like this:

1. Determine what information is being presented and what is being asked.

2. Formulate all possible hypotheses.

3. Check the consequences of each hypotheses against the data (the given information).

4. Reject all hypotheses that are incompatible with the data. Retain all hypotheses that are compatible with the data.

5. If no hypothesis is compatible with the data, return to step 1.

CHAPTER

5

Linkage I: Basic Eukaryotic Chromosome Mapping

Important Terms and Concepts

Recombination is the process that generates gametes with gene combinations different from that seen in the parental source. Recombination can be **interchromosomal**, involving genes on nonhomologous chromosomes, or **intrachromosomal**, involving homologous chromosomes.

If the recombination is interchromosomal, an equal frequency of all gamete types results. If the recombination is intrachromosomal, the frequency of gamete types depends on the physical distance between the genes being studied. Genes that are located more than 50 map units from each other are said to be **unlinked**; all gametes occur in equal frequency. Genes that are 50 map units or fewer apart are said to be **linked**; the frequency of parental type gametes will be greater than 50 percent and the frequency of recombinant type gametes will be less than 50 percent.

Recombination of linked genes results from a physical exchange of genetic material between homologous chromosomes. The **chiasmata** seen in meiosis I are considered to be the cytological evidence of **crossing-over**, the process that leads to recombination. The percentage of recombination is used as a measure of the physical distance between two genes. It is calculated by the following formula:

$$\frac{(100\%)\,(\text{number of recombinant progeny})}{\text{total number of progeny}} = \text{number of map units (m.u.)}$$

Linkage maps are formed by combining the results from a series of crosses.

A **three-point testcross** can be done to determine the relative location of three genes simultaneously. To do a three-point testcross, a triple-recessive parent is mated with a triple-heterozygous parent. The contribution of the testcross parent is ignored. This is because the parent must contribute the three recessive alleles to each of the progeny.

The **gene order** is determined by comparing the parental type progeny, which will be most frequent, with the **double-crossover** progeny, which will be the least frequent. The gene that has switched with respect to the other two genes in this comparison is the gene in the middle:

parentals: $D\ a\ R$, $d\ A\ r$

double crossovers: $d\ a\ R$, $D\ A\ r$

The order is $A\ D\ R$; the parental chromosomes are $a\ D\ R/A\ d\ r$.

Map units in a three-point testcross are determined by identifying progeny that result from a crossover in each region. In the example above, single crossovers in the A–D region would be $a\ d\ r$ and $A\ D\ R$. Single crossovers in the D–R region would be $a\ D\ r$ and $A\ d\ R$. Because double-crossover progeny result from a crossover in both regions, their frequency is added to the frequency of single crossovers for each region. The formula for calculating the distance between two genes is

$$\frac{(100\%)\,(\text{number of single crossovers} + \text{number of doublecrossovers})}{\text{total number of progeny}}$$
$$= \text{number of map units}$$

The expected types and numbers of progeny can be determined from a linkage map. To determine the expected number of double crossovers, multiply the total number of progeny by the map units in both regions. To determine the expected number of single crossovers, multiply the total number of progeny by the map units in that region and subtract the number of double crossovers from the total.

Interference (I) occurs when a crossover in one region affects the frequency of a crossover in a second region. It is defined as

$$1 - \frac{\text{observed double crossovers}}{\text{expected double crossovers}}$$

If interference exists, it needs to be calculated from data when determining map distances or used to adjust frequencies when calculating the expected progeny types from a linkage map. In the latter situation, the value of I will be given. To calculate the observed number of double crossovers, use the following formula:

$$\text{observed double crossovers} = (1 - \text{I})\,(\text{expected double crossovers})$$

Use the **chi square** test to decide whether observations are compatible with the hypothesis that generated the expected values.

Be sure that you have thoroughly read the entire chapter before you attempt any of the problems.

Solutions to Problems

1. The *Aa Bb* progeny are parentals; 90 percent of the progeny are parentals, and they are of two types: *Aa Bb* and *aa bb*. Therefore, 45 percent of the progeny will be *Aa Bb*.

2.

P	$A\ d/A\ d \times a\ D/a\ D$
F_1	$A\ d/a\ D$
F_2	$1\ A\ d/A\ d$
	$2\ A\ d/a\ D$
	$1\ a\ D/a\ D$

3.

P	$R\ S/r\ s \times R\ S/r\ s$
gametes	$1/2\ (1 - 0.35)\ R\,S$
	$1/2\ (1 - 0.35)\ r\,s$
	$1/2\ (0.35)\ R\,s$
	$1/2\ (0.35)\ r\,S$

F_2	0.1056 *R S/R S*
	0.1056 *r s/r s*
	0.2113 *R S/r s*
	0.1138 *R S/r S*
	0.1138 *R S/R s*
	0.1138 *r s/r S*
	0.1138 *r s/R s*
	0.0306 *R s/R s*
	0.0306 *R s/R s*
	0.0613 *R s/r S*

4. The cross is *Ee Ff* × *ee ff*. If independent assortment exists, the progeny should be in a 1:1:1:1 ratio, which is not observed. Therefore, there is linkage. *E f* and *e F* are recombinants equaling one-third of the progeny. The two genes are 33.3 map units (m.u.) apart.

5. There may not be enough genes identified in the animal to make it probable that two would be identified on the same chromosome within 50 m.u. Also, there may be a very high frequency of crossing-over.

6. Because only parental types were recovered, the two genes must be quite close to each other, making recombination quite rare.

7. Parental types are the most frequent: 442 *Aa Bb* and 458 *aa bb*. Because one parent was *aa bb*, contributing only *a b* to the offspring, the parental types can be rewritten as 442 *A B/a b* and 458 *a b/a b*. Thus, the female parent was *A B/a b*. The two recombinant types are 46 *A b/a b* and 54 *a B/a b*. The frequency of recombination between two genes is

$$\frac{100\%\,(\text{total number or recombinants})}{\text{total number of progeny}}$$

$$= \frac{100\%\,(46+54)}{(442+458+46+54)} = \frac{100\%\,(100)}{1000} = 10\text{ m.u.}$$

8. Meiosis is occurring in an organism that is *C d/c D*, producing haploid spores ultimately. The parental types are *C d* and *c D*, in equal frequency. The recombinant types are *C D* and *c d*, in equal frequency. Eight map units means 8 percent recombinants. Thus, *C D* and *c d* will each be present at a frequency of 4 percent, and *C d* and *c D* will each be present at a frequency of (100% – 8%)/2 = 46%.

4 percent; **b.** 4 percent; **c.** 46 percent; **d.** 8 percent

9. To solve this problem, you must realize that

1. One chiasma involves two of the four chromatids in a homologous pair. Therefore, 16 percent of the meioses having a chiasma will lead to 8 percent recombinants.

2. One-half of the recombinants will be *B r* and one-half will be *b R*. The answer is **b**, 4 percent.

10. a. Gene pairs *A/a*, *B/b*, and *C/c* are linked, and *D/d* shows no recombination with *A*/a. This is determined by looking at only the two genes you are trying to make a decision about.

Are *A/a* and *B/b* linked?

$$A\ B = 140 + 305 = 445$$
$$a\ b = 145 + 310 = 455$$
$$a\ B = 42 + 6 = 48$$
$$A\ b = 43 + 9 = 52$$

The two genes are 10 m.u. apart.

Are *A/a* and *D/d* linked?

$$A\ D = 0$$
$$a\ d = 0$$
$$A\ d = 43 + 140 + 9 + 305 = 497$$
$$a\ D = 42 + 145 + 6 + 310 = 503$$

The two genes show no recombination = 0 m.u..

Are *B/b* and *D/d* linked?

$$B\ D = 42 + 6 = 48$$
$$b\ d = 43 + 9 = 52$$
$$B\ d = 140 + 305 = 445$$
$$b\ D = 145 + 310 = 455$$

The two genes are 10 m.u. apart.

Are *C/c* and *D/d* linked?

$$C\ D = 42 + 310 = 350$$
$$c\ d = 43 + 305 = 348$$
$$C\ d = 140 + 9 = 149$$

$$cD = 145 + 6 = 151$$

The two genes are 30 m.u. apart.

All four genes are linked.

b. and **c.** Because A/a and D/d show no recombination, first rewrite the progeny omitting D and d (or omitting A and a).

$a\ B\ C$	42
$A\ b\ c$	43
$A\ B\ C$	140
$a\ b\ c$	145
$a\ B\ c$	6
$A\ b\ C$	9
$A\ B\ c$	305
$a\ b\ C$	310
	1000

Note that the progeny now look like a typical three-point testcross, with $A\ B\ c$ and $a\ b\ C$ the parental types (most frequent) and $a\ B\ c$ and $A\ b\ C$ the double recombinants (least frequent). The gene order is $B/b\ A/a\ C/c$. This is determined by comparing double recombinants with the parentals; the gene that "switches" in reference with the other two is the gene in the center ($B\ A\ c \rightarrow B\ a\ c$, $b\ a\ C \rightarrow b\ A\ C$).

Next, rewrite the progeny again, this time putting the genes in the proper order, and classify the progeny.

$B\ a\ C$	42	co A–B
$b\ A\ c$	43	co A–B
$B\ A\ C$	140	co A–C
$b\ a\ c$	145	co A–C
$B\ a\ c$	6	(DCO)
$b\ A\ C$	9	(DCO)
$B\ A\ c$	305	parental
$b\ a\ C$	310	parental

To calculate the map of these genes, use the following formula:

distance between two genes

$$= \frac{100\% \text{ (number of single co + number of DCO)}}{\text{total number of progeny}}$$

For the A/a to B/b distance:

$$\frac{100\%\,(42 + 43 + 6 + 9)}{1000} = \frac{100\%\,(100)}{1000} = 10 \text{ m.u.}$$

For the A/a to C/c distance:

$$\frac{100\%\,(140 + 145 + 6 + 9)}{1000} = \frac{100\%\,(300)}{1000} = 30 \text{ m.u.}$$

The map is B 10 m.u. A 30 m.u. C

Now it is time to deal with the D/d alleles. Notice that only two combinations were observed with A/a: $A\ d$ and $a\ D$. The parental chromosomes actually were $B\ (A, d)\ c/b\ (a,D)\ C$, where the parentheses indicate that the order of the genes within is unknown.

d. Interference = 1 – [(observed DCO)/(expected DCO)]

$= 1 - \{(6+9)/[(0.10)(0.30)(1000)]\}$

$= 1 - (15/30) = 0.5$

11. a. Males must be heterozygous for both genes, and the two must be closely linked: $M\ F/m\ f$.

b. $m\ f/m\ f$

c. Sex is determined by the male contribution. The two parental gametes are $M\ F$, determining maleness ($M\ F/m\ f$), and $m\ f$, determining femaleness ($m\ f/m\ f$). Occasional recombination would yield $M\ f$, determining a hermaphrodite ($M\ f/m\ f$), and $m\ F$, determining total sterility ($m\ F/m\ f$).

d. Recombination in the male yielding $M\ f$

e. Hermaphrodites are rare because the genes are tightly linked.

12. The verbal description indicates the following cross and result:

P $\quad N\!-\ A\!-\times nn\ OO$

F_1 $\quad Nn\ AO \times Nn\ AO$

The results indicate linkage, so the cross and results must be rewritten:

P	$N\,A/?\,? \times n\,O/n\,O$
F_1	$N\,A/n\,O \times N\,A/n\,O$
F_2	66% $N\text{–}\,A\text{–}$
	16% $n\,O/n\,O$
	9% $nn\,A\text{–}$
	9% $N\,O/?\,0$

Only one genotype is fully known: 16% $n\,O/n\,O$, a combination of two parental gametes. The frequency of two parental gametes coming together is the frequency of the first times the frequency of the second. Therefore, the frequency of each $n\,O$ is the square root of 0.16, or 0.4. Within an organism the two parental gametes occur in equal frequency. Therefore, the frequency of $N\,A$ is also 0.4. The parental total is 0.8, leaving 0.2 for all recombinants. Therefore, $N\,O$ and $n\,A$ occur at a frequency of 0.1 each. The two genes are 20 m.u. apart.

13. P yellow female × scarlet male

F_1 251 wild-type females: therefore, two genes are involved and both are recessive defects

248 yellow males: therefore, the defect in the female line is X-linked

Let *d* stand for the defective gene in the female and *e* for the defective gene in the male. The cross becomes

P $dd\ EE \times DY\ e\text{–}$, where the dash stands for either *e* or the Y chromosome

F_1 $Dd\ Ee$ wild-type female

$dY\ E\text{–}$ yellow males

If E/e is on an autosome, the male is $dY\ Ee$ and the F_1 cross should result in

females	males
9 $D\text{–}\ E\text{–}$ wild-type	3 $DY\ E\text{–}$ wild-type
3 $dd\ E\text{–}$ yellow	1 $DY\ ee$ scarlet
3 $D\text{–}\ ee$ scarlet	3 $dY\ E\text{–}$ yellow

1 *dd ee* 1 *dY ee*

Because the ratios are not close to what was observed, the defect in the male parent must also be X-linked. The cross needs to be rewritten again:

P	$d\ E/d\ E \times D\ e/Y$	
F_1	$d\ E/D\ e \times d\ E/Y$	
F_2	females	males
	$d\ E/d\ E$ yellow (parental)	179 $d\ E/Y$ yellow (parental)
	$D\ e/d\ E$ wild-type (parental)	183 $D\ e/Y$ scarlet (parental)
	$D\ E/d\ E$ wild-type (co)	77 $D\ E/Y$ wild-type (co)
	$d\ e/d\ E$ yellow (co)	80 $d\ e\ Y$ brown (co)

The data from the female are ignored because parentals cannot be distinguished from recombinants. Wild-type and brown in the males are recombinants. The recombination frequency is

$$100\%\ (77 + 80)/(179 + 183 + 77 + 80) = 30.25 \text{ m.u.}$$

14. Because of the unequal frequency among the progeny, the two genes must be linked. The parental chromosomes were $ADH^F PGM^S/ADH^S PGM^F$. The recombination frequency is

$$100\%\ (19 + 21)/(19 + 21 + 95 + 102) = 16.88 \text{ m.u.}$$

15. The original cross was

P $P\ L/P\ L \times p\ l/p\ l$

F_1 $P\ L/p\ l \times P\ L/p\ l$

Before proceeding, recognize that crossing-over can occur in both parents and that some crossovers cannot be detected by phenotype. The gametes from each plant are as follows:

parental types: $P\ L$, $p\ l$

recombinants: $P\ l$, $p\ L$

The F_2 is as follows:

$P\ L/P\ L$ purple, long	$p\ l/p\ L$ red, long
$P\ L/p\ l$ purple, long	$p\ l/p\ l$ red, round

$P\,L/P\,l$ purple, long	$P\,l/p\,L$ purple, long
$P\,L/p\,L$ purple, long	$P\,l/P\,l$ purple, round
$P\,l/p\,l$ purple, round	$p\,L/p\,L$ red, long

Of these ten different genotypes, only red, round can be identified unambiguously. It consists of two parental type gametes and occurs at a frequency of 14.4 percent. The probability of such a genotype can be calculated by multiplying the probability of a gamete from the first parent times the probability of the same gamete from the second parent. Thus, the square root of 14.4 percent will yield the frequency of this one type of parental gamete, or 37.99 percent. Because parental types occur with equal frequency, the parentals are 75.98 percent and the recombinants are 24.02 percent. Therefore, there are approximately 24 m.u. between the two genes.

A more precise calculation involves the use of an advanced statistical technique known as the method of maximum likelihood.

16. P $a\ b\ c/a\ b\ c \times a^+b^+c^+/a^+b^+c^+$

F_1 $a\ b\ c/a^+b^+c^+ \times a\ b\ c/a^+b^+c^+$

F_2 1364 $a^+\text{–}\ b^+\text{–}\ c^+\text{–}$

365 $a\ b\ c/a\ b\ c$

87 $aa\ bb\ c^+\text{–}$

84 $a^+\text{–}\ b^+\text{–}\ cc$

47 $aa\ b^+\text{–}\ c^+\text{–}$

44 $a^+\text{–}\ bb\ cc$

5 $aa\ b^+\text{–}\ cc$

4 $a^+\text{–}\ bb\ c^+\text{–}$

Remember that recombination does not occur in the male *Drosophila*.

a. Because you cannot distinguish between $a\ b\ c/a^+\ b^+\ c^+$ and $a^+\ b^+\ c^+/a^+\ b^+\ c^+$, use the frequency of $a\ b\ c/a\ b\ c$ to estimate the frequency of $a^+\ b^+\ c^+$ gametes from the female.

parentals	730	(2×365)
co *a–b*:	91	$(a++, +\,b\,c = 47 + 44)$
co *b–c*:	171	$(a\,b+, ++\,c = 87 + 84)$
DCO:	9	$(a+c, +\,b+ = 5 + 4)$
	1001	

a–b: 100% $(91+9)/1001 = 10$ m.u.

b–c: 100% $(171+9)/1001 = 18$ m.u.

b. Coefficient of coincidence = (observed DCO)/(expected DCO)

= 9/[(0.1)(0.18)(1001)]

= 9/18 = 0.5

17. a. By comparing the two most frequent classes (parentals: *an* br^+ f^+, an^+ *br f*) to the least frequent classes (DCO: an^+ *br* f^+, *an* br^+ *f*), the gene order can be determined. The gene in the middle switches with respect to the other two (the order is *an f br*). Now the crosses can be written fully.

P $an\ f^+\ br^+/an\ f^+\ br^+ \times an^+\ f\ br/an^+\ f\ br$

F_1 $an^+\ f\ br/an\ f^+\ br^+ \times an\ f\ br/an\ f\ br$

F_2 355 $an\ f\ br/an\ f^+\ br^+$, parental

339 $an\ f\ br/an^+\ f\ br$, parental

88 $an\ f\ br/an^+\ f^+\ br^+$, co *an–f*

55 $an\ f\ br/an\ f\ br$, co *an–f*

21 $an\ f\ br/an^+\ f\ br^+$, co *f–br*

17 $an\ f\ br/an\ f^+\ br$, co *f–br*

2 $an\ f\ br/an^+\ f^+\ br$, DCO

2 $an\ f\ br/an\ f\ br^+$, DCO

b. *an–f*: 100%(88 + 55 + 2 + 2)/879 = 16.72 m.u.

f–br: 100%(21 + 17 + 2 + 2)/879 = 4.78 m.u.

an 16.72 m.u. f 4.78 m.u. br

c. Interference = 1 – [(observed DCO)/(expected DCO)]

= 1 – {4/[(0.1672)(0.0478)(879)]}

= 1 – 0.569 = 0.431

18. By comparing the most frequent classes (parental: + *v lg*, *b* + +) with the least frequent classes (DCO: + + +, *b v lg*) the gene order can be determined. The gene in the middle switches with respect to the other two, yielding the following sequence: *v b lg*. Now the cross can be written

P $v\ b^+ lg/v^+\ b\ lg^+ \times v\ b\ lg/v\ b\ lg$

F_1 305 $v\ b\ lg/v\ b^+\ lg$, parental

275 $v\ b\ lg/v^+\ b\ lg^+$, parental

128 $v\ b\ lg/v^+\ b\ lg$, co b–lg

112 $v\ b\ lg/v\ b^+\ lg^+$, co b–lg

74 $v\ b\ lg/v^+\ b^+\ lg$, co v–b

66 $v\ b\ lg/v\ b\ lg^+$, co v–b

22 $v\ b\ lg/v^+\ b^+\ lg^+$, DCO

18 $v\ b\ lg/\ v\ b\ lg$, DCO

v–b: 100%(74 + 66 + 22 + 18)/1000 = 18.0 m.u.

b–lg: 100%(128 + 112 + 22 + 18)/1000 = 28.0 m.u.

c.c. = (observed DCO)/(expected DCO)

= (22 + 18)/[(0.28)(0.18)(1000)] = 0.79

19. The F_1 males indicate that rickets and hemophilia are X-linked. The F_2 males also indicate that tail-less is X-linked. The F_1 females indicate that all three genes are recessive. The gene order is determined by comparing in the male the most frequent classes (parentals: tail-less versus hemophilia and rickets) with the least frequent classes (DCO: rickets versus tail-less and hemophilia). The gene order is tail-less hemophilia rickets. The cross is

a. P $T\ h\ r/T\ h\ r \times t\ H\ R/Y$

F_1 $T\ h\ r/t\ H\ R \times T\ h\ r/Y$

F_2 males

437 $t\ H\ R/Y$	parental
439 $T\ h\ r/Y$	parental
48 $T\ H\ R/Y$	co T–H
46 $t\ h\ r/Y$	co T–H
12 $T\ h\ R/Y$	co H–R
12 $t\ H\ r/Y$	co H–R
4 $T\ H\ r/Y$	DCO
2 $t\ h\ R/Y$	DCO

T–H: 100%(48 + 46 + 4 + 2)/1000 = 10 m.u.

H–R: 100%(12 + 12 + 4 + 2)/1000 = 3.0 m.u.

b. Interference = 1 – [(observed DCO)/(expected DCO)]

= 1 – {6/[(0.1)(0.03)(1000)]} = –1.0

20. Let F = fat, L = long tail, and Fl = flagella. The gene sequence is F L Fl (compare most frequent to least frequent). The cross is

P $F\ L\ Fl/f\ l\ fl \times f\ l\ fl/f\ l\ fl$

F_1 398 $F\ L\ Fl/f\ l\ fl$, parental

370 $f\ l\ fl/f\ l\ fl$, parental

72 $F\ L\ fl/f\ l\ fl$, co L–Fl

67 $f\ l\ Fl/f\ l\ fl$, co L–Fl

44 $f\ L\ FL/f\ l\ fl$, co F–L

35 $F\ l\ fl/f\ l\ fl$, co F–L

9 $f\ L\ fl/f\ l\ fl$, DCO

5 $F\ l\ Fl/f\ l\ fl$, DC0

L–Fl: 100%(72 + 67 + 9 + 5)/1000 = 15.3 m.u.

F–L: 100%(44 + 35 + 9 + 5)/1000 = 9.3 m.u.

F 9.3 m.u. L 15.3 m.u. Fl

21.a–b. The data indicate that the progeny males have a different phenotype than the females. Therefore, all the genes are on the X chromosome. The two most frequent phenotypes in the males indicate the X chromosomes in the female, and the two least frequent phenotypes in the males indicate the gene order. The cross is

P $x\ z\ +/+\ +\ y \times +\ +\ +/Y$

F_1 males

430 $x\ z$ +/Y, parental

441 + + y/Y, parental

39 $x\ z\ y$/Y, co z–y

30 + + +/Y, co z–y

32 + z +/Y, co x–z

27 x + y/Y, co x–z

$1 + z\ y/\text{Y}$, DCO

$0\ x + +/\text{Y}$, DCO

c. z–y: $100\%(39 + 30 + 1)/1000 = 7.0$ m.u.

x–z: $100\%(32 + 27 + 1)/1000 = 6.0$ m.u.

c.c. = (observed DCO)/(expected DCO)

$= 1/[(0.06)(0.07)(1000)] = 0.238$

d. X chromosome

22. Because all F_1 flies are wild-type, the brown and humpy genes are not X-linked. Also, all wild-type indicates that all the deviations from wild-type are recessive. Because wild-type eye color results from a cross of scarlet with brown, two genes for eye color are involved. The scarlet phenotype shows up in the F_2 males, but not the females, indicating X-linkage for this gene. Let h = humpy, s = scarlet, b = brown. The cross is

P $hh\ bb\ SS \times HH\ BB\ s/\text{Y}$

F_1 $Hh\ Bb\ Ss \times Hh\ Bb\ S/\text{Y}$

If humpy is linked to brown, there will be a lack of independent assortment (not 9:3:3:1) for these two genes. Because humpy and brown are autosomal and scarlet is X-linked, scarlet will independently assort with respect to either humpy or brown. White eyes must be a combination of *bb s*/Y.

The F_2 male progeny are

145 wild-type	H– B– S/Y
139 wild thorax, scarlet	H– B– s/Y
40 humpy thorax, white eyes	$hh\ bb\ s/\text{Y}$
39 humpy thorax, brown eyes	$hh\ bb\ S/\text{Y}$
11 humpy thorax, scarlet eyes	hh B– s/Y
10 wild thorax, white eyes	H– $bb\ s/\text{Y}$
9 wild thorax, brown eyes	H– $bb\ S/\text{Y}$
8 humpy thorax, wild eyes	hh B– S/Y

The ratio of 284 *H– B–* : 19 *H– bb* : 19 *hh B–* : 79 *hh bb* is very far from a 9:3:3:1 ratio, indicating linkage. The F_1 cross needs to be rewritten

F_1 *hb/HB S/s* × *hb/HB S/Y*

Remember that crossing-over does not occur in male flies. The F_2 male progeny can now be rewritten

145 *H– B– S/*Y

139 *H– B– s/*Y

40 *h b/h b s/*Y, parental with regard to *H/h* and *B/b*

39 *h b/h b S/*Y, parental with regard to *H/h* and *B/b*

11 *h B/h b s/*Y, co *H–B*

10 *H b/h b s/*Y, co *H–B*

9 *H b/h b S/*Y, co *H–B*

8 *h B/h b S/*Y, co *H–B*

If all the genotypes are written out, it will become clear that one-half of the crossover progeny have a wild-type phenotype because the male donates *H B* (not humpy, not brown or white) to one-half the crossovers and the female donates *S* to one-half the crossover (not scarlet, not white).

Male crossover (only) genotypes and phenotypes:

*H b/H B S/*Y wild-type

*H b/H B s/*Y scarlet; cannot be distinguished from non-co scarlet

*H b/h b S/*Y brown

*H b/h b s/*Y white

*h B/H B S/*Y wild-type

*h B/H B s/*Y scarlet; cannot be distinguished from non-co scarlet

*h B/h b S/*Y humpy

*h B/h b s/*Y humpy, scarlet

Therefore, the actual number of crossover progeny is twice the 38 identified above.

The only parental types that can be identified are those that obtained *h* and *b* from boh parents, 40 + 39 = 79. However, each parent can be assumed to have produced *HB* and *h b* in equal proportion. This should result in 1 *HB/HB* : 2 *HB/h b* : 1 *h b/h b*. In other words, only one fourth of the parental types have been identified. There are 4 × 79 = 316 parental types, three-fourths of them within the *H– B–* phenotype. Therefore, the distance between *H* and *B* is

$$\frac{100\%(2)(11+10+9+8)}{(316+11+10+9+8)} = 19.387\ \text{m.u}$$

When an independent analysis is done in the female progeny, the *H B* distance is 20.0 m.u. If you do not understand this explanation, write out all genotypes.

23. a.

Rh^+ E (R– E–) — Rh^- e (rr ee)

Rh^- e (rr ee) — Rh^+ E (Rr Ee)

1: Rh^+ e (Rr ee) 4: Rh^+ E (Rr Ee) 5: Rh^- e (rr ee)

b. yes

c. dominant

d. As drawn, the pedigree indicates independent assortment. However, the data also support linkage, with the *R e/r e* individual representing a crossover. The distance between the two genes would be 100%(1/10) = 10 m.u. There is no way to choose between the alternatives without more data.

24. The cross is

P *P A R/P A R* × *p a r/p a r*

F_1 *P A R/p a r* × *p a r/p a r*, a three-point testcross

a. number of parentals = 1 – (single co individuals – DCO individuals) = 1 – {[0.15 + 0.20 – 2(0.15)(0.20)] – [(0.15)(0.20)]} = 0.68. Because one-half of the parentals are Earth alleles and one-half are Vulcan, the frequency of children with all three Vulcanian characteristics is 1/2(0.68) = 0.34.

b. same as above, 0.34.

c. The frequency will be one-half the DCOs, or 1/2(0.15)(0.20) = 0.015.

d. The frequency will be 1/2 p(co *P–A*) p(no co *A–R*) = 1/2(0.15)(0.80) = 0.06.

25. a. To obtain a plant that is *a b c/a b c* from selfing of *A b c/a B C*, both gametes must be derived from a single crossover between *A* and *B*. The frequency of the *a b c* gamete is

$$1/2\ p(\text{co } A\text{–}B)\ p(\text{no co } B\text{–}C) = 1/2(0.20)(0.70) = 0.07$$

The frequency of an *a b c / a b c* plant is $(0.07)^2 = 0.0049$.

b. The cross is *A b c/a B C* X *a b c/a b c*. The progeny are

A b c/a b c, parental: $1/2[1 - (\text{co } A\text{–}B) - (\text{co } B\text{–}C) + \text{DCO}](1000)$

$= 1/2\{1 - 0.20 - 0.30 + [(0.20)(0.30)]\}(1000)$

$= 280$

a B C/a b c, parental: $1/2[1 - (\text{co } A\text{–}B) - (\text{co } B\text{–}C) + \text{DCO}](1000)$

$= 1/2\{1 - 0.20 - 0.30 + [(0.20)(0.30)]\}(1{,}000)$

$= 280$

A B C/a b c, co *A–B*: $1/2[(\text{co } A\text{–}B) - \text{DCO}](1000)$

$= 1/2\{0.20 - [(0.20)(0.30)]\}(1000) = 70$

a b c/a b c, co *A–B*: $1/2[(\text{co } A\text{–}B) - \text{DCO}](1000)$

$= 1/2\{0.20 - [(0.20)(0.30)]\}(1000) = 70$

A b C/a b c, co *B–C*: $1/2[(\text{co } B\text{–}C) - \text{DCO}](1000)$

$= 1/2\{0.30 - [(0.20)(0.30)]\}(1000) = 120$

a B c/a b c, co *B–C*: $1/2[(\text{co } B\text{–}C) - \text{DCO}](1000)$

$= 1/2\{0.30 - [(0.20)(0.30)]\}(1000) = 120$

A B c/a b c, DCO: $1/2(\text{DCO})(1000)$

$= 1/2(0.20)(0.30)(1000) = 30$

a b C/a b c, DCO: $1/2(\text{DCO})(1000)$

$= 1/2(0.20)(0.30)(1000) = 30$

c. Interference = 1 – (observed DCO)/(expected DCO)

$$0.2 = 1 - (\text{observed DCO})/(0.20)(0.30)$$

$$\text{observed DCO} = (0.20)(0.30) - (0.20)(0.20)(0.30) = 0.048$$

Of 1000 progeny, 48 will be DCO. The progeny are

A b c/a b c, parental: $1/2[1 - (\text{co } A\text{–}B) - (\text{co } B\text{–}C) + \text{DCO}](1000)$

$= 1/2(1 - 0.20 - 0.30 + 0.048)(1000) = 274$

$a\ B\ C/a\ b\ c$, parental: $1/2[1 - (\text{co } A\text{–}B) - (\text{co } B\text{–}C) + \text{DCO}](1000)$

$$= 1/2(1 - 0.20 - 0.30 + 0.048)(1000) = 274$$

$A\ B\ C/a\ b\ c$, co A–B: $1/2[(\text{co } A\text{–}B) - \text{DCO}](1000)$

$$= 1/2(0.20 - 0.048)(1000) = 76$$

$a\ b\ c/a\ b\ c$, co A–B: $1/2[(\text{co } A\text{–}B) - \text{DCO}](1000)$

$$= 1/2(0.20 - 0.048)(1000) = 76$$

$A\ b\ C/a\ b\ c$, co B–C: $1/2[(\text{co } B\text{–}C) - \text{DCO}](1000)$

$$= 1/2(0.30 - 0.048)(1000) = 126$$

$a\ B\ c/a\ b\ c$, co B–C: $1/2[(\text{co } B\text{–}C) - \text{DCO}](1000)$

$$= 1/2(0.30 - 0.048)(1000) = 126$$

$A\ B\ c/a\ b\ c$, DCO: $1/2(\text{DCO})(1000) = 1/2(48) = 24$

$a\ b\ C/a\ b\ c$, DCO: $1/2(\text{DCO})(1000) = 1/2(48) = 24$

26. Assume there is no linkage. The genotypes should occur with equal frequency, which is the expected value. In each case, there are four genotypes ($n = 4$), which means there are 3 degrees of freedom ($n - 1 = 3$).

$$X^2 = \text{sum (observed} - \text{expected)}^2/\text{expected}$$

a. $$X^2 = \frac{(310 - 300)^2 + (315 - 300)^2 + (287 - 300)^2 + (288 - 300)^2}{300}$$

$$= \frac{(100 + 225 + 169 + 144)}{300} = 2.1266$$

$P > 0.50$, nonsignificant. Therefore, the hypothesis of no linkage cannot be rejected.

b. $$X^2 = \frac{(36 - 30)^2 + (38 - 30)^2 + (23 - 30)^2 + (23 - 30)^2}{30}$$

$$= \frac{36 + 64 + 49 + 49}{30} = 6.6$$

$P > 0.10$, nonsignificant. The hypothesis of no linkage cannot be rejected.

c. $$X^2 = \frac{(360 - 300)^2 + (380 - 300)^2 + (230 - 300)^2 + (230 - 300)^2}{300}$$

$$= \frac{3600 + 6400 + 4900 + 4900}{300} = 66.0$$

$P < 0.005$, significant. The hypothesis of no linkage must be rejected.

d. $$X^2 = \frac{(74 - 60)^2 + (72 - 60)^2 + (50 - 60)^2 + (44 - 60)^2}{60}$$

$$= \frac{196 + 144 + 100 + 256}{60} = 11.60$$

$P < 0.01$, significant. The hypothesis of no linkage must be rejected.

27. The data approximate a 9:3:3:1 ratio, which suggests two genes. Let A = resistance to rust 24, a = susceptibility to rust 24, B = resistance to rust 22, b = susceptibility to rust 22.

a.

P *AA bb* (770B) × *aa BB* (Bombay)

F_1 *Aa Bb* × *Aa Bb*

F_2 184 *A– B–*

58 *A– bb*

63 *aa B–*

15 *aa bb*

320

b. Expect:

(320) (9/16) = 180 *A– B–*

(320) (3/16) = 60 *A– bb*

(320) (3/16) = 60 *aa B–*

(320) (1/16) = 20 *aa bb*

$$X^2 = \frac{(184 - 180)^2}{180} + \frac{(58 - 60)^2}{60} + \frac{(63 - 60)^2}{60} + \frac{(15 - 20)^2}{20}$$

$$= \frac{16}{180} + \frac{4}{60} + \frac{9}{60} + \frac{25}{20} = 1.555$$

P (3 df) > 0.58, nonsignificant. A P value is the probability that the result would be observed by chance alone. Therefore, the hypothesis of two independently assorting genes cannot be rejected.

28. If h = hemophilia and b = color blindness, the genotypes for individuals in the pedigree can be written as

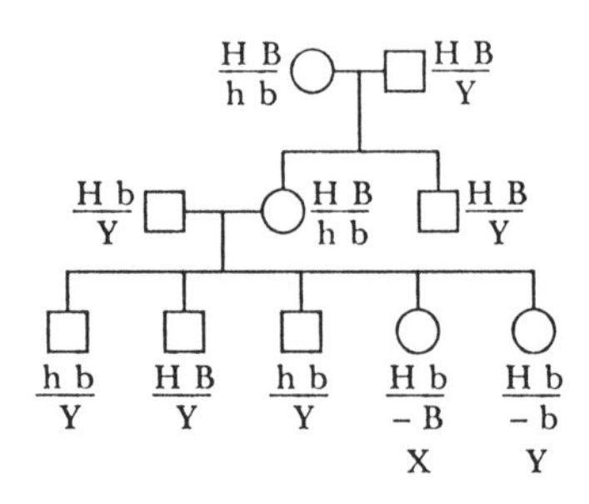

The mother of the two women in question would produce the following gametes:

0.45 *H B*	0.05 *H b*
0.45 *h b*	0.05 *h B*

Woman X can be either *H b*/*H B* (0.45 chance) or *H b*/*h B* (0.05 chance), because she received *B* from her mother. If she is *H b*/*h B* [0.05/(0.45 + 0.05) = 0.10 chance], she will produce the parental and recombinant gametes with the same probabilities as her mother. Thus, her child has a 45 percent chance of receiving *h B*, a 5 percent chance of receiving *h b*, and a 50 percent chance of receiving a Y from his father. The probability that her child will be a hemophiliac son is (0.1)(0.50)(0.5) = 0.025 = 2.5 percent.

Woman Y can be either *H b*/*H b* (0.05 chance) or *H b*/*h b* (0.45 chance), because she received *b* from her mother. If she is *H b*/*h b* [0.45/(0.45 + 0.05) = 0.90 chance], she has a 50 percent chance of passing *h* to her child, and there is a 50 percent chance that the child will be male. The probability that she will have a son with hemophilia is (0.9)(0.5)(0.5) = 0.225 = 22.5 percent.

29. a. Cross 1 reduces to

P *AA BB DD* × *aa bb dd*

F_1 *A B D*/*a b d* × *a b d*/*a b d*, correct order

F_2 *A B D* 316 parental

a b d 314 parental

A B d 31 co *B–D*

a b D 39 co *B–D*

A b d 130 co *A–B*

a B D 140 co *A–B*

A b D 17 DCO

a B d 13 DCO

A–B: 100%(130 + 140 + 17 + 13)/1000 = 30 m.u.
B–D: 100%(31 + 39 + 17 + 13)/1000 = 10 m.u.

Cross 2 reduces to

P *AA CC EE* × *aa cc ee*

F_1 *A C E*/*a c e* × *a c e*/*a c e*, correct order

F_2 *A C E* 243 parental

a c e 237 parental

A c e 62 co *A–C*

a C E 58 co *A–C*

A C e 155 co *C–E*

a c E 165 co *C–E*

a C e 46 DCO

A c E 34 DCO

A–C: 100%(62 + 58 + 46 + 34)/1000 = 20 m.u.
C–E: 100%(155 + 165 + 46 + 34)/1000 = 40 m.u.

The map can be put together in one way that accommodates all the data: E 40 m.u. C 20 m.u. A 30 m.u. B 10 m.u. D

b. Interference (*I*) = 1 – [(observed DCO)/(expected DCO)]

For cross 1:

$I = 1 - \{30/[(0.30)(0.10)(1000)]\} = 1 - 1 = 0$, no interference.

For cross 2:

$I = 1 - \{80/[(0.20)(0.40)(1000)]\} = 1 - 1 = 0$, no interference.

Tips on Problem Solving

In a three-point testcross, the gene order is determined by comparing the parental type progeny, which will be most frequent, with the double-crossover progeny, which will be least frequent. The gene that has switched with respect to the other two genes in this comparison is the gene in the middle.

parentals:	*D a R, d A r*
double crossovers:	*d a R, D A r*

The order is *A D R*; the parental chromosomes are *a D R*/*A d r* (see Problem 18).

In three-point testcrosses, if the data indicate that the progeny males have a different phenotype than the females, the genes are on the X chromosome. The two most frequent phenotypes in the males indicate the X chromosomes in the female, and the two least frequent phenotypes in the males indicate the gene order (see Problems 21, 13). To estimate the frequency of crossing-over, utilize the data from the males only (see Problem 19).

If, in a three-point cross, the data indicate that the genes are autosomal, it is frequently necessary to focus on the homozygous recessive progeny exclusively (see Problem 15). Also, if the cross is not a testcross, it may be necessary to use the homozygote recessive frequency to estimate the frequency of crossing-over (see Problem 16).

Self-Test

1. Given the following map, what are the recombinant genotypes and their frequency from the cross *M n*/*m N* × *m n*/*m n*?

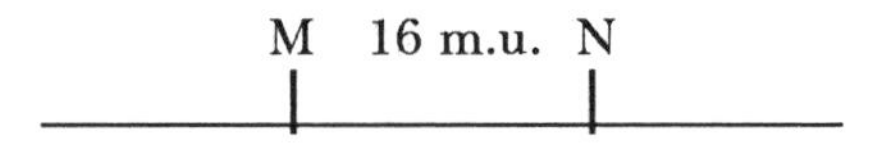

2. There is 24 percent crossing-over in humans between two autosomal genes, *L* and *M*. In 250 primary oocytes, how many would be expected to have a chiasma between these two genes?

3. In *Drosophila*, a cross was made between two homozygous lines, A and B. Line A females carried a mutant allele that resulted in rough eye texture. They were crossed with line B males, who had a mutant allele for a different gene that resulted in rough eye texture. The F_1 progeny

were wild-type females and males with a rough eye texture. The F_2 progeny were

females	males
50% wild-type	13% wild-type
50% rough	87% rough

Analyze these data.

4. Consider the following F_2 progeny.

0.41 *Dd Ee*	0.045 *dd Ee*
0.2025 *dd EE*	0.045 *Dd ee*
0.2025 *DD ee*	0.0025 *DD EE*
0.045 *Dd EE*	0.0025 *dd ee*
0.045 *DD Ee*	

Are the genes linked? If linked, by how many m.u.? What were the parental chromosomes?

5. A testcross resulted in the following progeny. Construct a gene map from the data. What is the interference value?

375 *Aa Bb dd*	90 *Aa Bb Dd*
380 *aa bb Dd*	85 *aa bb dd*
26 *aa Bb Dd*	8 *aa Bb dd*
30 *Aa bb dd*	6 *Aa bb Dd*

6. You are given the following map and the information that interference is 0.4. What is the change in the number of double-crossover progeny due to interference?

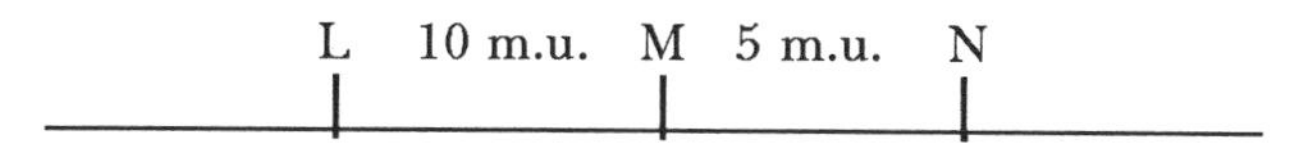

7. In rats, a mutant recessive allele, *k*, results in kinky hairs. Another mutant recessive allele, *t*, results in a short, stubby tail. The two genes are 30 m.u. apart. What are the progeny phenotypes from two heterozygous parents, in coupling?

Solutions to Self-Test

1. The recombinants will be *M N*/*m n* and *m n*/*m n*. Each will occur at a rate of 1/2(0.16) = 8% percent.

2. Recall that each recombination event gives rise to two recombinants and two parentals. Therefore, the frequency of chiasmata is (0.24)(250) = 60 oocytes with a chiasma.

3. The F_1 progeny indicate that the defect in the female line (line A) is X-linked. If the line B defect is autosomal, it would result in independent assortment in the F_2 male progeny. This was not observed. Therefore, both genes are X-linked.

Call the defect in line A *rr* and the defect in line B *ss*. The cross is

P	*r S*/*r S* × *R s*/Y	
F_1	*r S*/*R s* × *r S*/Y	
F_2	males	
	r S/Y	rough
	R s/Y	rough
	r s/Y	rough
	R S/Y	wild-type

The *R S*/Y genotype occurred in 13 percent of the population. It occurred in 50 percent of the recombinants. Therefore, the two genes are 26 m.u. apart.

4. If the genes are not linked, then a 9:3:3:1 ratio would occur. The observed ratio is

0.5025 *D– E–*	0.2475 *D– ee*
0.2475 *dd E–*	0.0025 *dd ee*

This ratio is very far from the expected ratio, assuming independent assortment. Therefore, the genes are linked.

The parental chromosomes can be determined from the least frequent classes, which represent crossovers. If crossovers are *DD EE* and *dd ee*, the F_1 chromosomes must be *D e*/*d E*. Assuming that the parentals were homozygous, the original cross must have been *DD ee* × *dd EE*.

The two least frequent classes each represent the fusion of recombinant gametes. The probability of such a fusion is the probability of a recombinant gamete times the probability of a recombinant gamete.

Therefore, the probability of a recombinant gamete is the square root of either genotypic class, or 0.05. Because each recombinant gamete has a reciprocal gamete occurring at the same frequency, the total frequency of recombination is (2)(0.05) = 0.10. Therefore, the genes are 10 m.u. apart.

5. The gene order is *B A D*, and the two parental chromosomes are *B A d*/*b a d*.

$$B\text{–}A\text{: } 100\%(26 + 30 + 8 + 6)/1000 = 7 \text{ m.u..}$$

$$A\text{–}D\text{: } 100\%(90 + 85 + 8 + 6)/1000 = 18.9 \text{ m.u.}$$

$$I = 1 - \{(8 + 6)/[(0.07)(0.189)(1000)]\} = 1 - 1.06 = -0.06.$$

6. The observed and expected number of double-crossovers, without interference, is (0.1)(0.05) = 0.005.

With interference of 0.4, the observed number of double-crossovers is

$$I = 1 - [(\text{observed DCO})/(\text{expected DCO})]$$

$$\begin{aligned} \text{observed DCO} &= \text{expected DCO} - [I(\text{expected DCO})] \\ &= 0.005 - [(0.4)(0.005)] \\ &= 0.003 \end{aligned}$$

The change in the number of double-crossovers is 0.002.

7. P *K T*/*k t* × *K T*/*k t*

To work this problem, first determine the gametic frequencies:

0.35 *K T*	0.15 *K t*
0.35 *k t*	0.15 *k T*

Most of the progeny will be wild-type. The remainder will be

Phenotype	Genotypes	Frequency
kinky	*kk TT* + *kk Tt*	(0.15)(0.15) + 2(0.35)(0.15) = 0.1275
short tail	*KK tt* + *Kk tt*	(0.15)(0.15) + 2(0.35)(0.15) = 0.1275
kinky, short tail	*kk tt*	(0.35)(0.35) = 0.1225
wild-type	*K– T–*	1 – 0.1275 – 0.1275 – 0.1225 = 0.6225

CHAPTER

6

Linkage II: Special Eukaryotic Chromosome Mapping Techniques

Important Terms and Concepts

Multiple crossovers in larger intervals result in an underestimate of map distance between two genes. The relationship between real map distance and the **recombination frequency**, RF, is not linear. The **mapping function** provides a closer approximation of the real relationship. The **Poisson distribution** describes the frequency of 0, 1, 2, . . . n crossovers, given the average number of crossovers. The general expression is

$$f(i) = e^{-m}m^{i}/i!$$

where e is the base of natural logarithms, m is the mean number of events, and i is an integer that ranges from 0 to n.

Meiotic events in diploid organisms are studied by means of an estimation based on the observation of random meiotic products. An indirect calculation of the recombination frequency is conducted using the Poisson distribution. In contrast, the four products of a single meiotic event can be studied directly through **tetrad analysis** in certain haploid organisms. This allows for a direct calculation of the recombination frequency, again using the Poisson distribution.

Linear tetrad analysis is frequently conducted using *Neurospora.* Recall that meiosis, followed immediately by mitosis, results in eight meiotic products.

A monohybrid cross in which no crossing-over occurs between the gene studied and the centromere results in **first-division segregation** of alleles and a 4:4 pattern of the progeny.

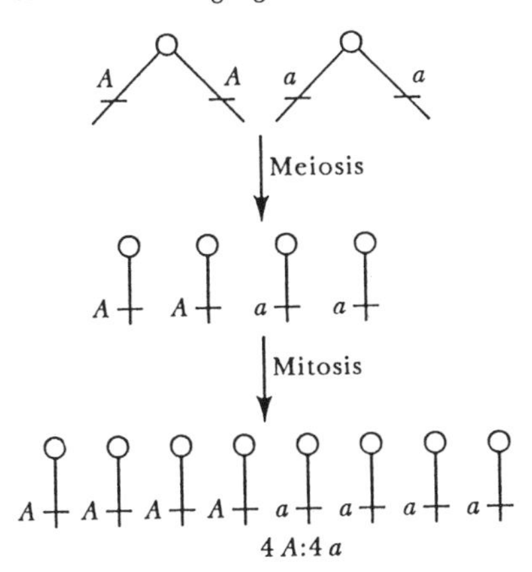

Any deviation from this pattern is the result of **second-division segregation** of alleles and signals crossing-over (shown on top of p. 100).

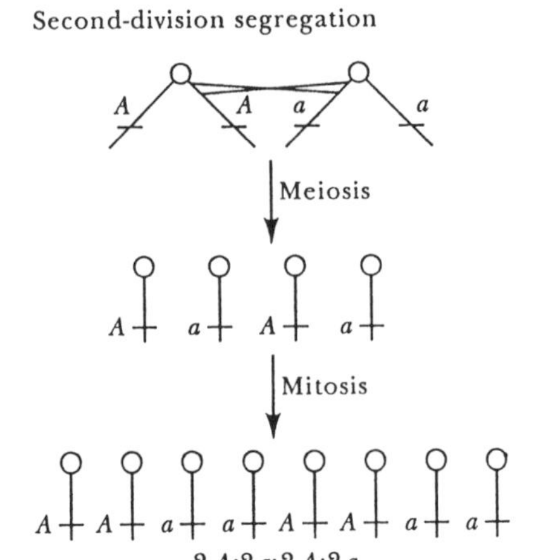

Because each crossover produces two recombinant and two nonrecombinant products, the frequency of recombinant asci must be adjusted to reflect the true RF value. The distance of a locus from the centromere is therefore

$$\mathrm{RF} = \frac{1/2 \text{ number of recombinant asci}}{\text{total number of asci}}$$

First-division segregation of two loci on different chromosomes results in two types of equally frequent 4:4 patterns, **parental ditype**, PD, and **nonparental ditype**, NPD. Parental ditype means that there are two types of spores and each spore is parental in allelic content. Nonparental ditype means that each spore is recombinant and there are two types of spores. For example,

Parentals: $A\ B \times a\ b$

PD: $A\ B, A\ B, A\ B, A\ B, a\ b, a\ b, a\ b, a\ b$

$4\ A\ B : 4\ a\ b$

NPD: $A\ b, A\ b, A\ b, A\ b, a\ B, a\ B, a\ B, a\ B$

$4\ A\ b : 4\ a\ B$

Any deviation from the two 4:4 patterns results from second-division segregation, signaling crossing-over between a gene and the centromere. One example is *A b*, *A b*, *A B*, *A B*, *a b*, *a b*, *a B*, *a B*, or 2:2:2:2. The asci that contain the allelic combinations *A B*, *a b*, *A b*, and *a B* are called **tetratypes**, T, indicating that there are four different types of meiotic products. A tetratype in unlinked loci indicates that crossing-over has occurred between a gene and its centromere. Whether or not crossing-over occurs, the number of parental ditypes equals the number of nonparental ditypes, and the NPD/T ratio is between 1/4 and infinity when the two genes are not linked. The RF between a gene and its centromere is calculated separately for each gene, using the formula

$$\mathrm{RF} = \frac{1/2\ \text{number of recombinant asci}}{\text{total number of asci}}$$

The number of parental ditypes significantly exceeds the number of nonparental ditypes when there are two linked loci. The NPD/T ratio is between 0 and 1/4. No crossing-over leads to a 4:4 pattern, while a single crossover between a gene and its centromere results in a tetratype. Two crossovers result in a parental ditype, a tetratype, or a nonparental ditype, depending on which strands are involved. The RF between a gene and its centromere is calculated separately for each gene, using the formula

$$\mathrm{RF} = \frac{1/2\ \text{number of recombinant asci}}{\text{total number of asci}}$$

The RF between the two genes is calculated using the following formula:

$$\mathrm{RF} = \frac{100\%\,[1/2\mathrm{T} + \mathrm{NPD}]}{\text{total number of asci}}$$

Because two crossovers can lead to a parental ditype, this formula results in an underestimate of the RF.

Mitotic crossing-over leading to recombination and **mitotic nondisjunction** can result in **mitotic segregation** of alleles following mitosis. Mitotic segregation has been extensively studied in fungal cells. A **heterokaryon** generated by hyphal fusion in *Aspergillus* consists of two nuclei in a common cytoplasm. Phenotypic variegation indicates mitotic segregation in the heterokaryons.

Human chromosomes have been mapped using **somatic cell hybridization**. Fusion of two nuclei, one human and the other frequently mouse, results in a gradual and random loss of human chromosomes. The presence or absence of human enzymes is correlated with the presence or absence of human chromosomes in several cell lines.

Be sure that you have thoroughly read the entire chapter before you attempt any of the problems.

Solutions to Problems

1. a. + +, *al–2 al–2*, + +, *al–2 al–2*; *al–2 al–2*, + +, *al–2 al–2*, + +

b. The 8 percent value can be used to calculate the distance between the gene and the centromere.

2. a. *arg–6 al–2* and + +

b. *arg–6* +, *arg–6 al–2*, + +, and + *al–2*

c. *arg–6* + and + *al–2*

3. The formula for this problem is $f(i) = e^{-m}m^i/i!$, where $m = 2$ and $i = 0, 1,$ and 2.

a. $f(0) = e^{-2}2^0/0! = e^{-2} = 0.135$

b. $f(1) = e^{-2}2^1/1! = e^{-2}(2) = 0.27$

c. $f(2) = e^{-2}2^2/2! = e^{-2}(2) = 0.27$

4. a. A region 1 crossover will yield the following chromosomes:

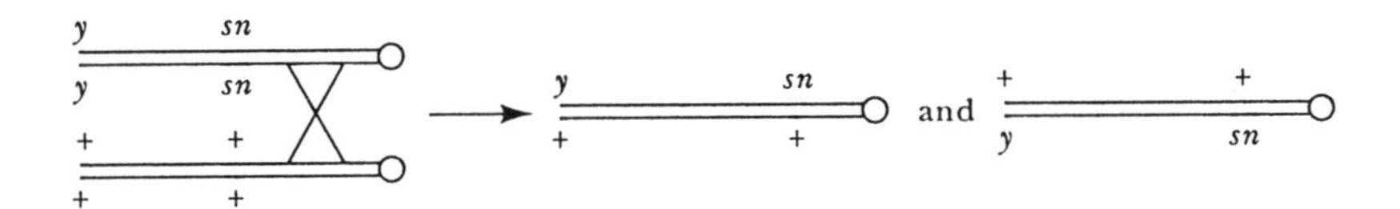

Mitosis can yield two equally likely results: *y sn*/*y sn* and + +/+ + or both daughter cells *y sn*/+ +. In the first case, one yellow singed spot will exist in a brown unsinged fly. In the second case, the resulting cells will be wild–type.

b. A region 2 crossover will yield the following chromosomes:

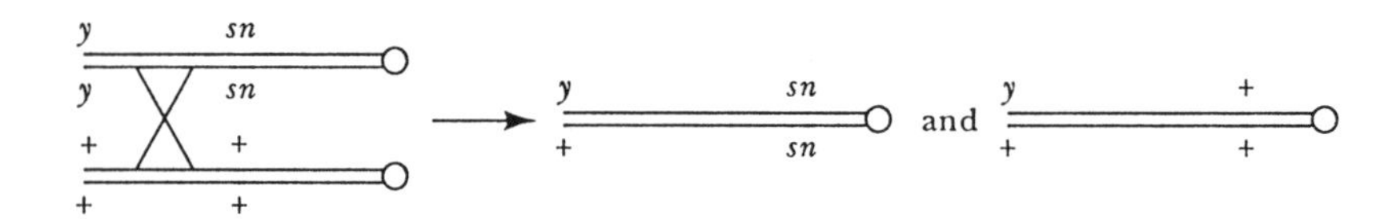

Mitosis can yield the following combinations:

y sn/+ + and + *sn* /*y* + wild-type

or

y sn/ *y* + and + *sn*/+ + yellow unsinged spot

5. Rewrite the column headings to note what is missing from the media, and then count the different types of patterns. Growth will occur if the wild-type gene is present or if the medium supplies whatever gene product is missing.

—*leu*	—*nic*	—*ad*	—*arg*	Number
+	+	–	–	6
–	–	+	+	4
–	+	–	+	5
+	–	+	–	4
+	–	–	–	1
+	–	–	+	0
–	+	+	–	0

This is a 1:1:1:1 pattern, indicating independent assortment of two chromosomes. Remember that the cross is

$$arg^- \ ad^- \ nic^+ \ leu^+ \times arg^+ \ ad^+ \ nic^- \ leu^-.$$

Growth is not seen in the (—nic and —ad) or (—leu and —arg) media simultaneously, which means that *nic* is linked to *ad* and *leu* is linked to *arg*. Both *nic* and *ad* assort independently with *leu* and *arg*, suggesting that *nic* is not linked to *leu* and *arg* and that *ad* is not linked to *leu* and *arg*.

a. The parents were

$$\underline{ad^- \ nic^+} \ \ \underline{leu^+ \ arg^-} \times \underline{ad^+ \ nic^-} \ \ \underline{leu^- \ arg^+}$$

b. Culture 16 resulted from a crossover between *ad* and *nic*. The reciprocal did not show up in the small sample.

6. This problem is analogous to meiosis in organisms that form linear tetrads. Let red = *R* and blue = *r*. Then meiosis is occurring in an organism that is *Rr* (but there are 4 "alleles" because the "chromosomes" are at the "two-chromatids-per-chromosome" stage), and the "alleles" are at loci far from the centromere. The patterns, their frequencies, and the division of segregation are given below. Notice that the probabilities change as each ball/allele is selected. This occurs when there is sampling without replacement.

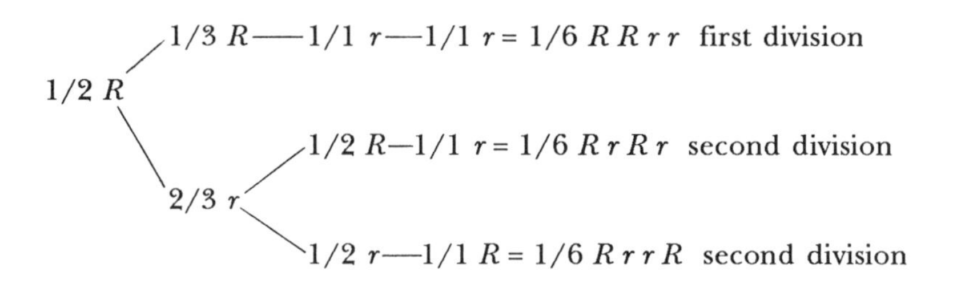

1/3 r—1/1 R—1/1 R = 1/6 r r R R first division
1/2 r
1/2 R—1/1 r = 1/6 r R R r second division
2/3 R
1/2 r—1/1 R = 1/6 r R r R second division

These results indicate one-third first division segregation and two-thirds second division segregation.

7. a. The formula is RF = $1/2(1 - e^{-m})$, where recombination = 0.2. Therefore, $e^{-m} = 1 - 0.4 = 0.6$, and $m = 0.51$ (from e^{-m} tables). Because an m value of 1.0 = 50 map units (m.u.), 0.51 × 50 m.u. = 25.5 m.u.

b. The problem is the interpretation of 45 m.u. That could represent two loci approximately 45 m.u. apart, or it could represent two unlinked loci. A X^2 test is needed for decision making. Hypothesis: no linkage, resulting in a 1:1:1:1 ratio.

$$X^2 = \frac{(58-50)^2 + (52-50)^2 + (47-50)^2 + (43-50)^2}{50}$$

$$= \frac{(64 + 4 + 9 + 49)}{50} = 2.52$$

With 3 degrees of freedom, the probability is greater than 10% that the genes are not linked. Therefore, the hypothesis of no linkage can be accepted.

8. a. The probability of the number of exchanges uses the formula $f(i) = e^{-m}m^i/i!$. The answers for a mean of 1.0 crossovers are found on page 122 of *An Introduction to Genetic Analysis*, 4th.ed.

The proportion that will result in an M_{II} pattern remains constant for all means. Thus the task is to determine what it is for any given mean. Consider Figure 6-3, which shows the outcome for zero, one and two crossovers. Obviously, zero crossovers lead to zero M_{II} meioses, one crossover leads to an M_{II} pattern, and two crossovers leads to M_I and M_{II} patterns in equal frequency. By drawing similar pictures for three and four crossovers, you will discover that 75 percent and 63 percent, respectively, of the meioses will be M_{II}. Alternatively, notice that each additional crossover changes an M_I into an M_{II} and changes one half of the M_{II} into an M_I.

The proportion of ascii showing M_{II} is the probability of the meiotic exchange multiplied by the proportion of meioses that result in M_{II}.

b. The following table summarizes all results.

	Number of exchanges					
	0	1	2	3	4	Total
m = 1	0.37	0.37	0.18	0.06	0.02	
	0	1	0.5	0.75	0.63	
	0	0.37	0.09	0.045	0.0126	0.52
m = 0.5	0.61	0.30	0.08	0.01	0	
	0	1	0.5	0.75	0.63	
	0	0.3	0.04	0.0075	0	0.35
m = 2	0.14	0.27	0.27	0.18	0.09	
	0	1	0.5	0.75	0.63	
	0	0.27	0.135	0.135	0.0567	0.60
m = 4	0.02	0.07	0.15	0.02	0.20	
	0	1	0.5	0.75	0.63	
	0	0.07	0.075	0.15	0.126	0.42

These results are graphed below.

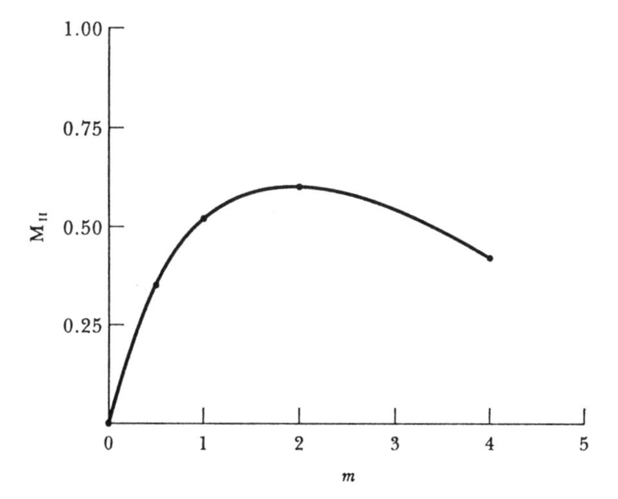

c. The curve bends downward because multiple crossovers are not detected as such. Practically, one corrects for this by studying genes that are rather close together to eliminate the probability of more than two crossovers between them.

9. To work these problems, it is first necessary to draw the chromosomes and explore the consequences of crossing-over in different regions. The genes can be assumed to be coupled or in repulsion.

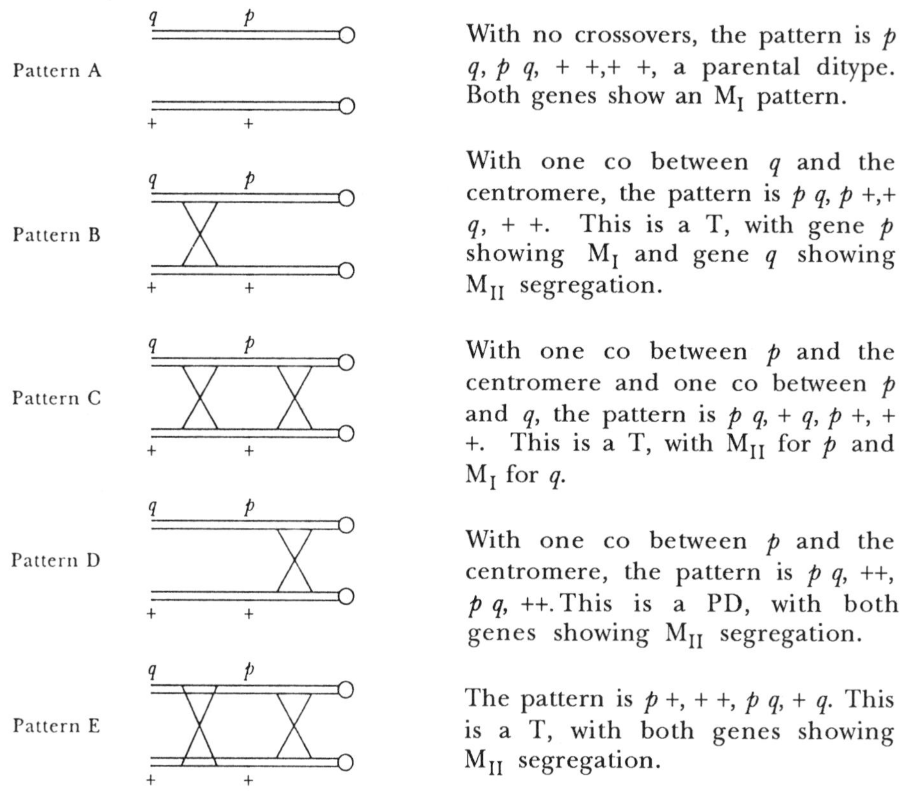

With no crossovers, the pattern is *p q*, *p q*, + +,+ +, a parental ditype. Both genes show an M_I pattern.

With one co between *q* and the centromere, the pattern is *p q*, *p* +,+ *q*, + +. This is a T, with gene *p* showing M_I and gene *q* showing M_{II} segregation.

With one co between *p* and the centromere and one co between *p* and *q*, the pattern is *p q*, + *q*, *p* +, + +. This is a T, with M_{II} for *p* and M_I for *q*.

With one co between *p* and the centromere, the pattern is *p q*, ++, *p q*, ++. This is a PD, with both genes showing M_{II} segregation.

The pattern is *p* +, + +, *p q*, + *q*. This is a T, with both genes showing M_{II} segregation.

a. M_I M_I, PD is pattern A, no crossovers. The probability = p(no co from centromere to *p*)p(no co from centromere to *q*) = (0.88)(0.80) = 0.704.

b. M_I M_I, NPD requires two crossovers between *q* and the centromere, which cannot occur according to the rules of the problem. The probability is 0.

c. M_I M_{II}, T is pattern B. The probability = p(no co from *p* to centromere)p(co between *q* and centromere) = (0.88)(0.2) = 0.176.

d. M_{II} M_I, T is pattern C. The probability = p(co between *p* and centromere)p(co between *q* and centromere) = (0.5)(0.12)(0.2) = 0.012.

e. M_{II} M_{II}, PD is pattern D. The probability = p(co between *p* and centromere)p(no co from *q* to centromere) = (0.12)(0.8) = 0.096.

f. M_{II} M_{II}, NPD requires one crossover between *p* and the centromere and two crossovers between *q* and the centromere, which cannot occur according to the rules of the problem. The probability is 0.

g. M_{II} M_{II}, T is pattern E. The probability = p(co between *p* and centromere)p(co between *q* and centromere) = (0.5)(0.12)(0.2) = 0.012.

10.

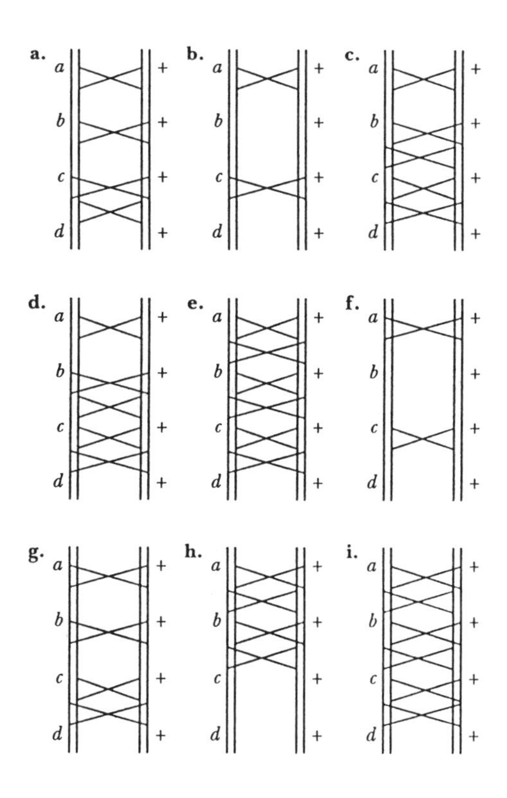

11. Before beginning this problem, classify all asci as PD, NPD, or T and determine whether there is M_I or M_{II} segregation for each gene.

	Asci Type						
	1	2	3	4	5	6	7
Type	PD	NPD	T	T	PD	NPD	T
gene *a* (M)	I	I	I	II	II	II	II
gene *b* (M)	I	I	II	I	II	II	II

If PD >> NPD and NPD/T $< 1/4$, there is linkage. If PD = NPD and NPD/T $> 1/4$, there is no linkage. The distance between a gene and its centromere = $100\%(1/2)(M_{II})/\text{total}$. The distance between two genes = $100\%(1/2T + NPD)/\text{total}$.

Cross 1: PD = NPD; the genes are not linked.

a—centromere: $100\%(1/2)(0)/100 = 0$ m.u. Gene *a* is very close to the centromere.

b—centromere: $100\%(1/2)(32)/100 = 16$ m.u.

O—— *a* ———— O———— *b* ——
0 16

Cross 2: PD >> NPD; the genes are linked.

a—*b*: $100\%[(1/2)(15) + 1]/100 = 8.5$ m.u.

a—centromere: $100\%(1/2)(0)/100 = 0$ m.u. Gene *a* is very close to the centromere.

b—centromere: $100\%(1/2)(15)/100 = 7.5$ m.u.

The data contradict because of asci that show an M_I pattern for *b* even though crossing-over must have occurred between *b* and the centromere. Therefore, the estimate of 8.5 m.u. between the two genes is the better estimate.

O—— *a* ———— *b* ——
0 8.5

Cross 3: PD >> NPD; the genes are linked.

a—*b*: $100\%[(1/2)(40) + (3)]/100 = 23$ m.u.

a—centromere: $100\%(1/2)(2)/100 = 1$ m.u.

b—centromere: $100\%(1/2)(40 + 2)/100 = 21$ m.u.

Again, the data are contradictory.

O—— *a* ———— *b* —— or —— *a* ——O———— *b* ——
1 23 1 21

The first diagram is the better interpretation of the data.

Cross 4: PD >> NPD; the genes are linked.

a—*b*: 100%[(1/2)(20) + 1]/100 = 11 m.u.

a—centromere: 100%(1/2)(10)/100 = 5 m.u.

b—centromere: 100%(1/2)(18 + 8 + 1)/100 = 13.5 m.u. Again, the *b*—centromere distance is underestimated.

O—— *a* ———— *b* ——
5 11

Cross 5: PD = NPD; the genes are not linked.

a—centromere: 100%(1/2)(22 + 8 + 10 + 20)/99 = 30.3 m.u.

b—centromere: 100%(1/2)(24 + 8 + 10 + 20)/99 = 31.3 m.u.

For values this large, the genes are considered unlinked to their centromeres in tetrad analysis.

Cross 6: PD >> NPD; the genes are linked.

a—*b*: 100%[(1/2)(1 + 3 + 4) + 0]/100 = 4 m.u.

a—centromere: 100%(1/2)(3 + 61 + 4)/100 = 34 m.u.

b—centromere: 100%(1/2)(1 + 61 + 4)/100 = 33 m.u.

The *b*-centromere distance is underestimated. Genes *a* and *b* are more than 50 m.u. from the centromere and are 4 m.u. apart.

O———— *a* —— *b* or O———— *b* —— *a*
>50 4 >50 4

Cross 7: PD >> NPD; the genes are linked.

a—*b*: 100%[(1/2)(3 + 2) + 0]/100 = 2.5 m.u.

a—centromere: 100%(1/2)(2)/100 = 1 m.u.

b—centromere: 100%(1/2)(3)/100 = 1.5 m.u.

—— *a* ——O———— *b* ——
1 1.5

Cross 8: PD = NPD; the genes are not linked.

a—centromere: 100%(1/2)(22 + 12 + 11+ 22)/100 = 33.5 m.u.

b—centromere: 100%(1/2)(20 + 12 + 11 + 22)/100 = 32.5 m.u.

Same as cross 5.

Cross 9: PD >> NPD; the genes are linked.

a—*b*: 100%[1/2(10 + 18 + 2) + 1]/100 = 16 m.u.

a—centromere: 100%(1/2)(18 + 1 + 2)/100 = 10.5 m.u.

b—centromere: 100%(1/2)(10 + 1 + 2)/100 = 6.5 m.u.

—— *a* ———O——— *b* ——
10.5 6.5

Cross 10: PD = NPD; the genes are not linked.

a—centromere: 100%(1/2)(60 + 1 + 2 + 5)/100 = 34 m.u.

b—centromere: 100%(1/2)(2 + 1 + 2 + 5)/100 = 5 m.u.

——O——— *a* — O——— *b* ——
>50 5

or

— *a* ———— *b* ———O
>50 5

or

— *a* ————O—— *b*
>50 5

Cross 11: PD = NPD; the genes are not linked.

a—centromere: 100%(1/2)(0)/100 = 0 m.u.

b—centromere: 100%(1/2)(0)/100 = 0 m.u.

O— *a* ——— O— *b* ———
0 0

12. The cross is *a* +/+ *b*, with the two genes unlinked. Each gene has a given probability of a crossover between it and the centromere. Remember that map units = 100%[(1/2)(number crossover asci)]/total asci. Therefore, if there are 5 m.u. between a gene and its centromere, there will be 10 percent crossover or second division asci (M_{II}) and 90 percent noncrossover or first division asci (M_I).

A 4:4 pattern exists with no crossing-over. Thus, if neither gene shows a crossover, two equally likely patterns are possible: *a b*, *a b*, ++, ++ (NPD) and *a* +, *a* +, + *b*, + *b* (PD).

If one gene (assume gene *a*) experiences a crossover, the pattern for it is *a*, +, *a*, + or *a*, +, +, *a*. Combined with a 4:4 pattern for gene *b*, this would result in four equally likely outcomes: *a b*, + *b*, *a* +, ++ (T) or *a b*, + *b*, ++, *a* + (T) or + *b*, *a b*, ++, *a* + (T) or *a b*, + *b*, ++, *a* + (T).

If both genes experience crossovers, the pattern is 1/4 PD : 1/2 T : 1/4 NPD (you should check to see if you can produce this ratio).

The following table presents the probabilities for both genes:

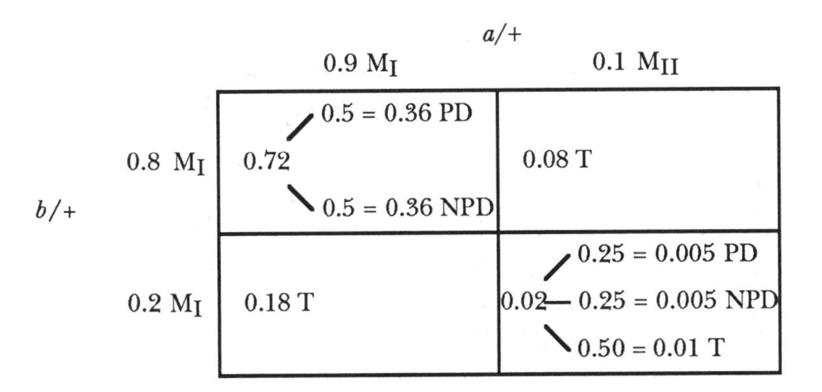

To understand the above table, look at cell 1. No crossovers occur between either gene and its centromere (p = 0.8 × 0.9). The pattern for *a* is thus *a a* + + and the pattern for *b* is *b b* + +. The *b*/+ gene can line up in either orientation with the *a*/+ gene: *a b a b* ++ ++ or *a* + *a* + + *b* + *b*. Therefore, there is a 1:1 chance of *a* + *a* + + *b* + *b* (PD) and *a b a b* ++ ++ (NPD). The other cells can be interpreted in a similar manner.

a. 36.5% PD: the total of PD from the table

b. 36.5% NPD: the total of NPD from the table

c. 27% T: the total of T from the table

d. 50% recombinants: 1/2 (T) + NPD

e. 25% growth in minimal medium: one-fourth of all spores will be + +

13. a. The pattern of crosses suggests that compatibility is determined by one locus with two alleles and that dikaryons can form between haploid colonies that carry differing allelic alternatives. Note that 1 and 3 can each cross with 2 and 4, but that 1 cannot cross with 3, and 2 cannot cross with 4. Let *m1* = allele in 1 and 3, *m2* = allele in 2 and 4. The tetrad was *m1*, *m2*, *m1*, *m2*.

b. Note that the pattern on each cultivar is 3:1. Also, recall that tetrads will give a 1:1 ratio for each gene.

Cultivar A: infected : uninfected = 3:1. This pattern suggests a dominant allele that results in infection. Call the allele *A*.

Cultivar B: infected : uninfected = 1:3. This pattern suggests a recessive allele that allows infection. Call the allele *b*, because it must be at a different locus than *A* (1 × 2 dikaryon did not infect cultivar A).

Cultivar C: infected : uninfected = 1:3. Again, a recessive allele is suggested, but it must be at a different locus than seen in cultivar B because the 1 × 2 dikaryon does not carry it. Call the allele *c*.

c. Begin by assigning genotypes to the dikaryons before attempting to assign genotypes to each haploid culture.

1 × 2: *aa bb C–*	3 × 2: *A– B– C–*
1 × 4: *A– B– C–*	3 × 4: *A– B– cc*

Immediately, you can write

1: *a b ?*	3: *? ? c*
2: *a b ?*	4: *? ? c*

Now compare the 1 × 4 cross with 3 × 4. Because 3 × 4 has a *cc* genotype, 4 must carry *c*, which means that 1 must carry *C*. The crosses can now be rewritten

1 × 2: *a b C/a b ?*

1 × 4: *a b C/A B c*

Because 2 is *a b ?*, the 3 × 2 cross indicates that 3 must be *A B c*. This also means that 2 must be *a b C*. The answer is

1: *a b C*	3: *A B c*
2: *a b C*	4: *A B c*

14. The frequency of recombinants is equal to NPD + 1/2T. The uncorrected map distance based on RF = (NPD + 1/2T)/total. The corrected map distance = 50(T + 6NPD)/total.

Cross 1:

recombinant frequency = 4% + 1/2(45%) = 26.5

uncorrected map distance = [4% + 1/2(45%)]/100% = 26.5 m.u.

corrected map distance = 50[45% + 6(4)]/100% = 34.5 m.u.

Cross 2:

recombinant frequency = 2% + 1/2(34%) = 19%

uncorrected map distance = [2% + 1/2(34%)]/100% = 19 m.u.

corrected map distance = 50[34% + 6(2%)]/100% = 29 m.u.

Cross 3:

recombinant frequency = 5% + 1/2(50%) = 30%

uncorrected map distance = [5% + 1/2(50%)]/100% = 30 m.u.

corrected map distance = 50[50% + 6(5%)]/100% = 40 m.u.

15. First, classify the asci: 138 T, 12 NPD, and 150 PD, for a total of 300 asci.

a. The frequency of recombinant asci is 50 percent, which leads to an uncorrected RF of 100%(12 + 69)/300 = 27 m.u. The corrected RF is m = –ln (1 – 0.54) = – ln(0.46) = 0.63, which is 31.5 m.u.

b. The general formula is DCOs = 4(NPD). Therefore, DCOs = 4(12) = 48. One half of them look like tetratypes and one half look like parental ditypes. Therefore

Actual 0 crossovers = PD – 12 = 150 – 12 = 138, 46%

Actual 1 crossovers = T – 24 = 138 – 24 = 114, 38%

Actual DCOs = 48, 16%

c. To correct for double crossovers, the general formula for the mean number of crossovers is m = T + 6NPD = [138 + 6(12)]/300 = 0.70, which is 35 m.u.

16. a. The cross is $arg^- \times arg^-$. Because arg^+ progeny result, more than one locus is involved, and each deviation from wild-type is recessive. The cross can be rewritten

$$arg\text{–}1^+\ arg\text{–}2^- \times arg\text{–}1^-\ arg\text{–}2^+$$

A 4:0 ascus is a PD ascus, because all spores require arginine. The 3:1 ascus must represent a T ascus. The 2:2 ascus is an NPD ascus.

b. The data support independent assortment of the two genes. PD = NPD = 40, and NPD/T = 40/20 = 2.00.

17. Because PD = NPD, the *his—?* is not linked to *ad*—3. The T—type ascospores require one crossover between the gene and the centromere.

Because only 10 tetrads were analyzed, *his—2* would be expected to produce one-tenth of a T ascus [10(0.01) = 0.1], *his—3* would be expected to produce one T ascus [10(0.1) = 1], and *his—4* would be expected to produce four T asci [10(0.4) = 4]. Only *his—4* is located far enough from its centromere to result in 60 percent T asci. Therefore, *his—?* is *his—4.*

18. Because the two mutants, when crossed, result in some black spores, two separate genes are involved, and both deviations from wild-type are recessive. Let mutant 1 = $w\ t^+$, mutant 2 = $w^+\ t$, and wild-type = w^+ t^+. The cross is

P $w\ t^+$ (white) $\times$ $w^+\ t$ (tan)

Asci types are

4 black : 4 white = 4 $w^+\ t^+$: 4 $w\ t$ (NPD)

4 tan : 4 white = 4 $w^+\ t$: 4 $w\ t^+$ (PD)

4 white : 2 black : 2 tan = 2 $w\ t^+$ (white) : 2 $w\ t$ (white):

2 $w^+\ t^+$: 2 $w^+\ t$ (T)

Notice that there are two types of white, indicating that w is epistatic to t.

19.

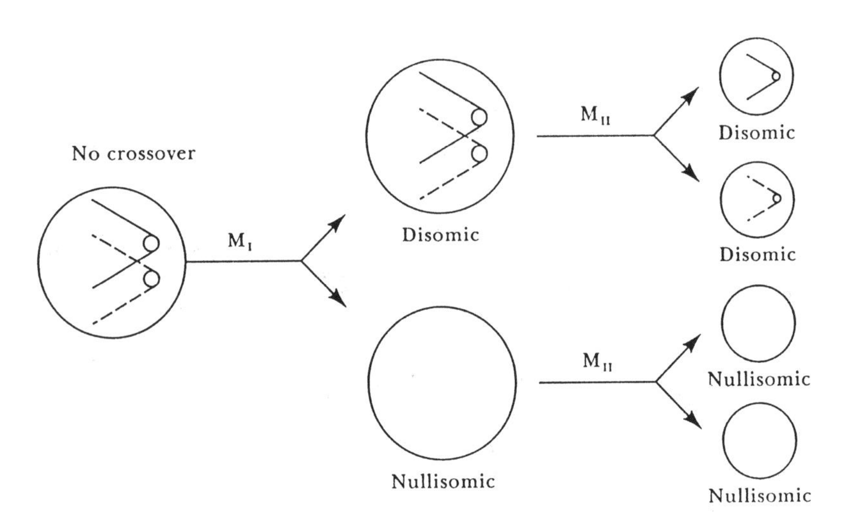

Egg	Sperm	Progeny
XX	X	XXX, nonviable
XX	Y	XXY, viable female
O	X	XO, sterile male
O	Y	OY, nonviable

a. A single crossover between *a* and *b*, involving strands 1 and 2 or involving strands 3 and 4, does not result in recombination. However, a 1—3 (shown below), 1—4, 2—3, or 2—4 crossover between *a* and *b* results in recombination.

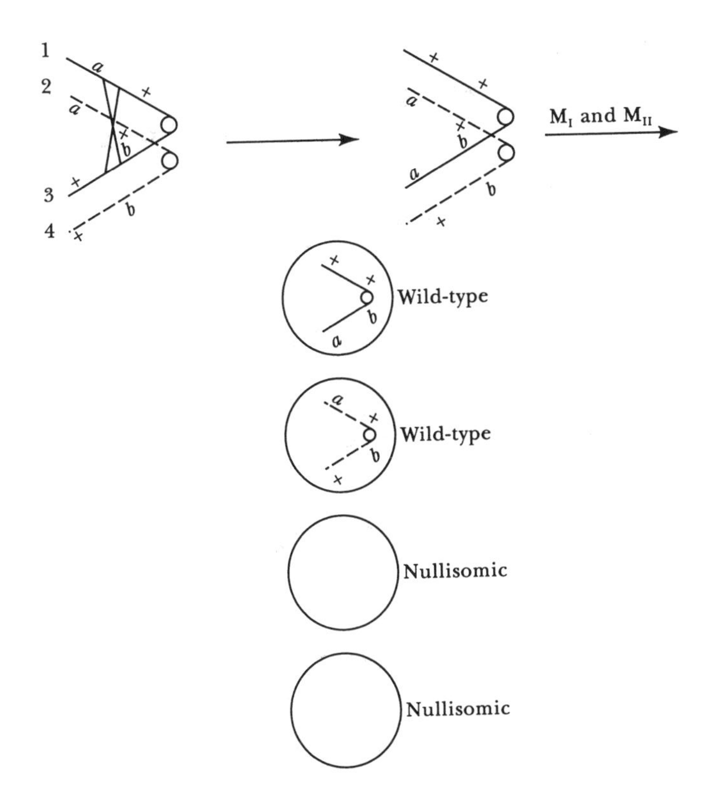

A single crossover between *b* and the centromere, involving strands 1 and 2 or involving strands 3 and 4, does not result in recombination. A 1—3 (shown on p. 116) or 2—4 crossover between *b* and the centromere also does not result in recombination.

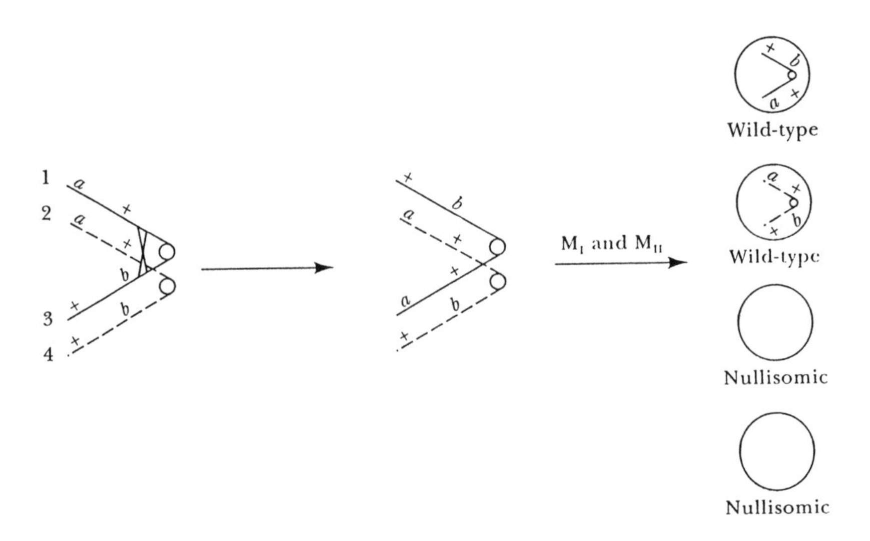

However, a 2—3 (illustrated below) or a 1—4 single crossover between *b* and the centromere does result in recombination.

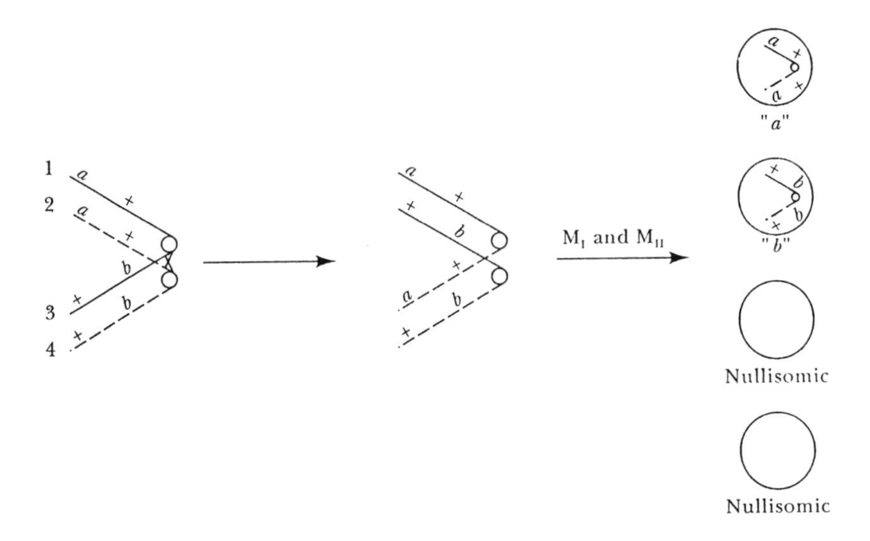

b. With double crossovers, the following are produced.

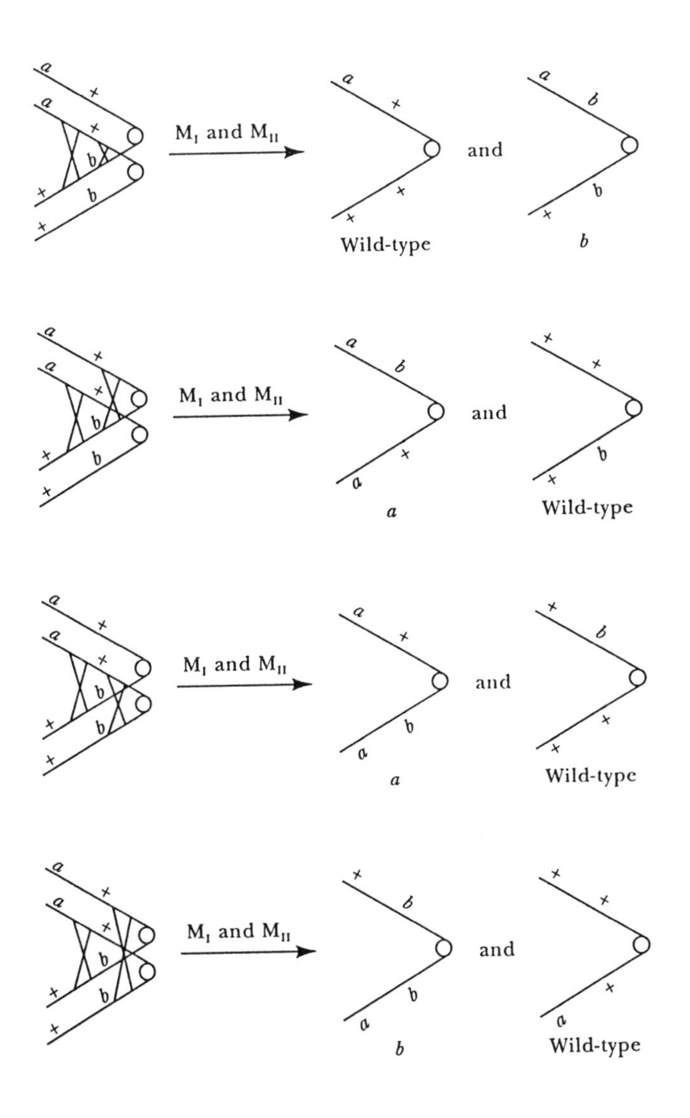

M$_{I}$ and M$_{II}$
and
Wild-type
b
M$_{I}$ and M$_{II}$
and
a
Wild-type
M$_{I}$ and M$_{II}$
and
a
Wild-type
M$_{I}$ and M$_{II}$
and
b
Wild-type

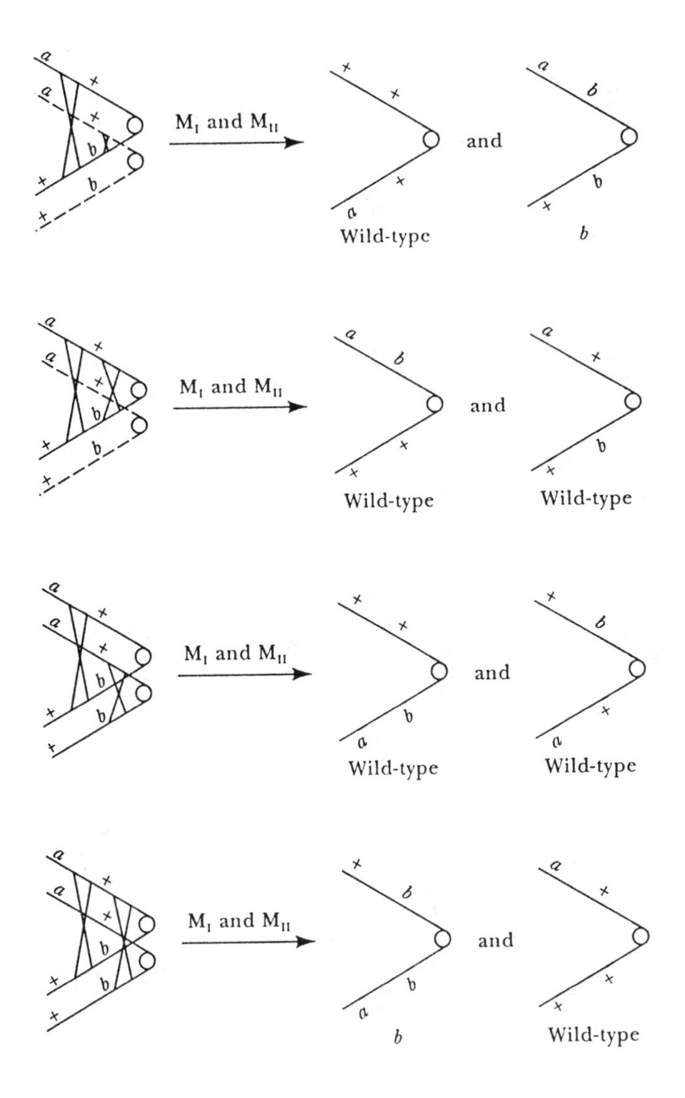

Note that the phenotype in each female (XXY) shows one-half the meiotic products. Thus, each female can be regarded as a half- tetrad.

20. Let *gg* = green, *Gg* = yellowish and *GG* = yellow. Mitotic crossing-over can account for the observations.

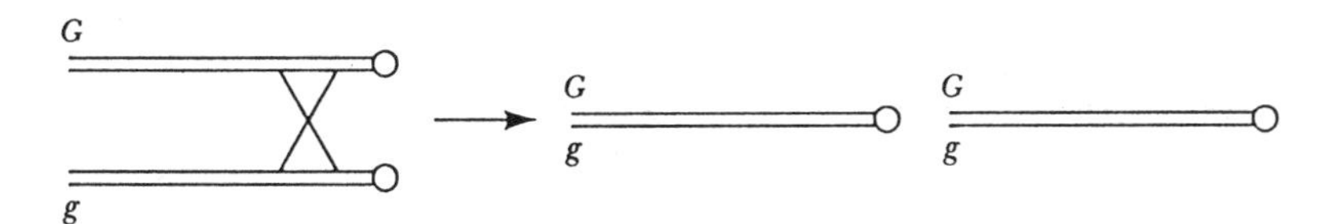

The resulting daughter cells would be *GG* (yellow) and *gg* (green).

21. The cross is $fw\ cha^+/fw^+\ cha$. The results suggest mitotic crossing-over.

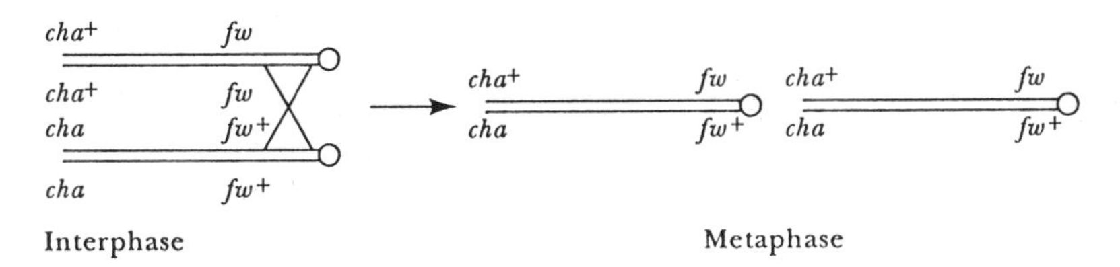

The resulting cells would be $cha^+\ fw/cha^+\ fw$ (fawn) and $cha\ fw^+/cha\ fw^+$ (chartreuse).

22. Ignoring the white allele for a moment, the 1:1:1:1 ratio indicates independent assortment of two units, or chromosomes. The *a* and *c* alleles are found in equal proportions with *b* and b^+, indicating that *b*/+ assorts independently of a *c*/+. If *w*/+ were linked on either chromosome, then the white allele would be found more frequently with one gene combination than another. Because it is not, *w*/+ is not linked to either chromosome.

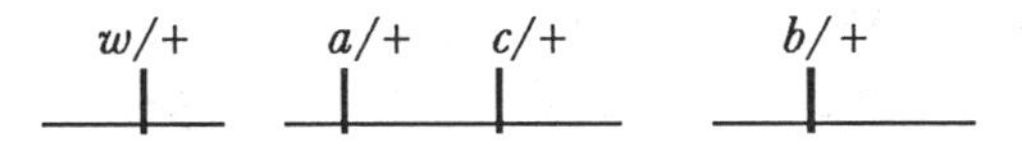

23. Remember that the segregants are yellow and diploid (80 percent $yy\ r^+-$, 20 percent $yy\ rr$). To get *yy* segregants, crossing-over must occur between *y* and the centromere. There are three possible arrangements:

1. yellow — centromere — ribo
2. centromere — yellow — ribo
3. centromere — ribo — yellow

If 1 is correct, after crossing-over the chromosomes would be

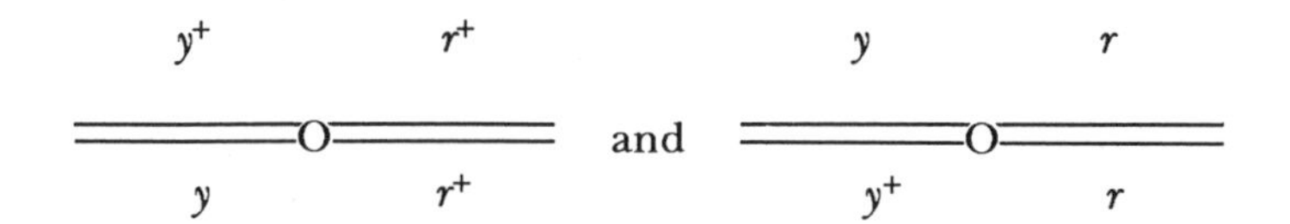

Without a crossover between r and the centromere, no rr segregants would be observed. Because the two genes are far away from each other, a 1:1 ratio of ribo-requiring and ribo-wild-type would be expected.

If 2 is correct, after crossing-over the chromosomes would be

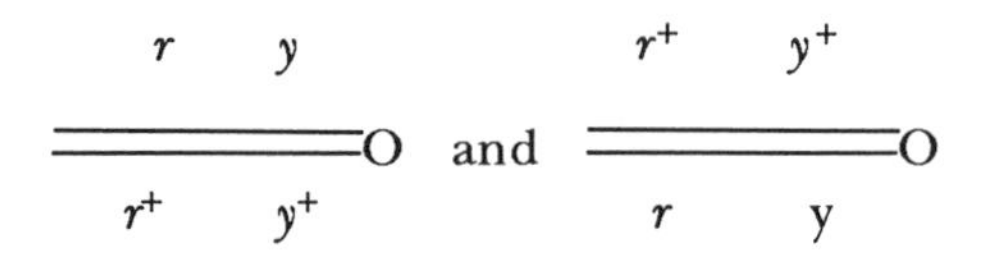

Unless the two genes are more than 50 m.u. apart, most of the yy segregants would be ribo-requiring.

If 3 is correct, crossing-over between r and the centromere would give ribo-requiring segregants, and crossing-over between y and the centromere would give ribo-wild-type. Therefore, both types of segregants would be observed with only one crossover. Furthermore, the 80:20 ratio suggests that crossing-over occurs more frequently between y and the centromere.

24. **a.** All *fpa/fpa* diploid segregants experienced a crossover between *fpa* and the centromere. The data indicate that *pro* and *paba* are linked to *fpa*. The different frequencies are a measure of the distances involved. Because there are no progeny requiring only *pro*, *pro* is closer to the centromere than *paba*.

b.

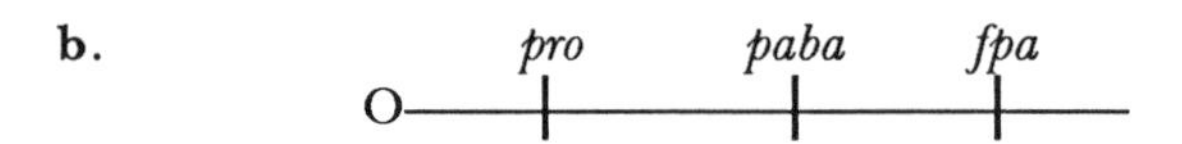

pro—paba: 100%(110)/154 = 71.4 relative m.u.
paba—fpa: 100%(35)/154 = 22.7 relative m.u.
pro—centromere: 100%(9)/154 = 5.8 relative m.u.

c. *pro paba fpa*

25. α: the only chromosome missing in A and B and present in C is 7.

ß: the only chromosome present in all colonies is 1.

γ: the only chromosome missing in A and present in B and C is 5.

Δ: the only chromosome present only in B is 6.

ε: not on chromosomes 1 through 7.

26. Steroid sulfatase: X*p*; Phosphoglucomutase—3: 6*q*; Esterase D: 13*q*; Phosphofructokinase: 21; Amylase: 1*p*; Galactokinase: 17*q*.

27.

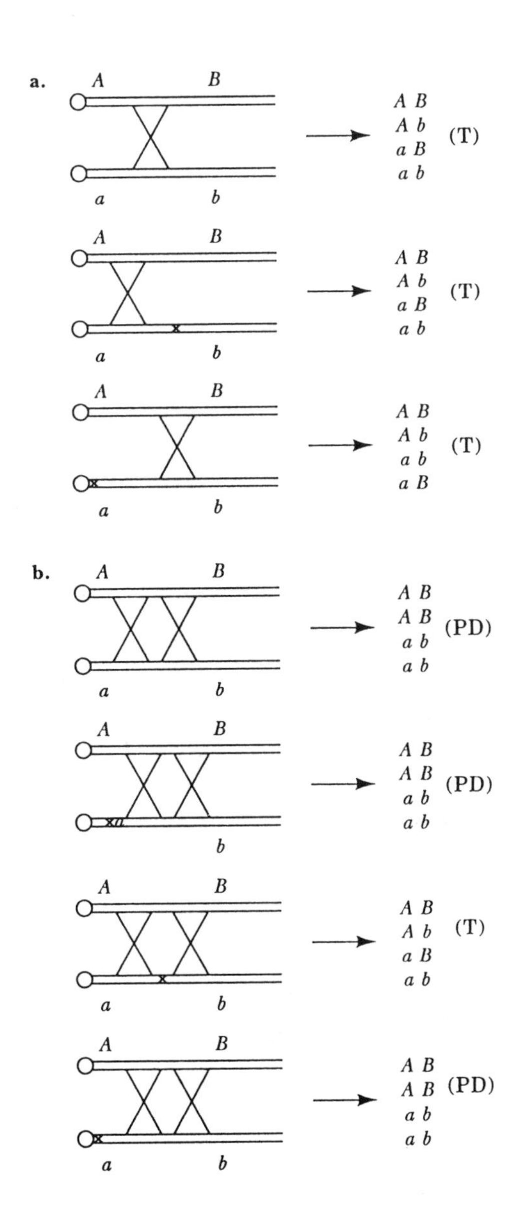

Three-strand double, type 1

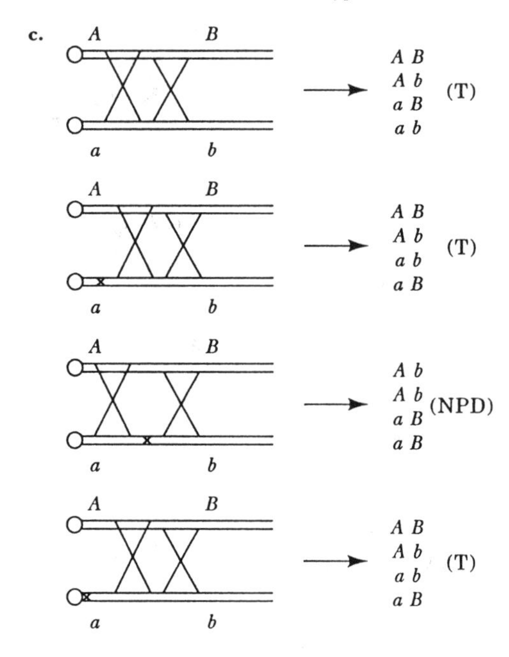

Three-strand double, type 2

A B

a b

A B
A b
a B
a b
(T)

A B

a b

A B
A b
a b
a B
(T)

A B

a b

A B
A B
a b
a b
(PD)

A B

a b

A B
A b
a b
a B
(T)

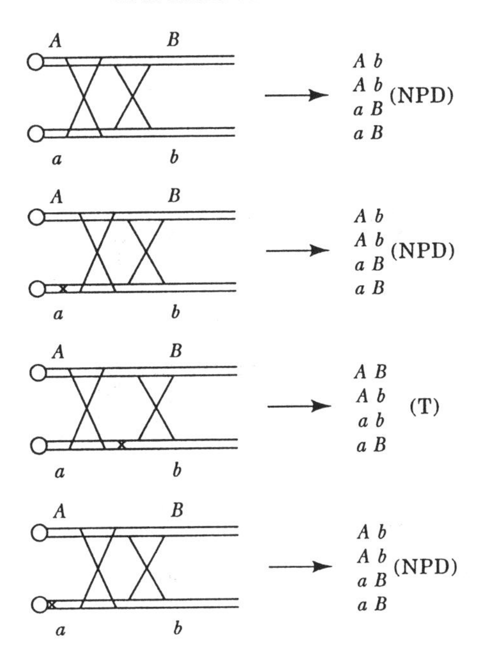

d.

Type of crossover	Effect of sister chromatid exchange
two-strand	PD → T
three-strand, type 1	T → NPD
three-strand, type 2	T → PD
four-strand	NPD → T

Assume that the two types of three-strand doubles occur with equal frequency. Then, each conversion listed in the table is equally likely to occur. Given this, the chance of a PD becoming a T through random sister chromatid exchange is the same as the chance of a T becoming a PD through sister chromatid exchange. The same holds true for the NPD to T conversions. Therefore, there is no net effect of sister chromatid exchange on the relative frequencies of PD, NPD, and T tetrad types.

Comments on This Material

The material covered in Chapter 6 of *An Introduction to Genetic Analysis*, 4th ed., will challenge the very best of students. Students experience more difficulty with this chapter than with any other chapter in the textbook. One reason for this may be insufficient mathematical skills, a problem that cannot be solved here. Another reason this material is so difficult for so many is that it requires the integration of all that you have learned up to this point. It requires a thorough understanding of the foregoing material and systematic application of the problem-solving techniques discussed earlier.

Self-Test

1. One strain (*ad*) of *Neurospora* requires adenine for growth. Another strain (*ylo*) produces yellow conidia. The two strains were crossed, and the following asci observed.

Number of asci	Ascus type
110	4 *ad* + : 4 + *ylo*
18	2 *ad* + : 2 *ad ylo* : 2 + + : 2 + *ylo*

a. Are the two genes linked?

b. Calculate the distance of each gene from its centromere and, if the genes are linked, the distance between the two genes.

2. Two strains of *Neurospora* requiring adenine were crossed and the following asci were observed.

Number of asci	Ascus type
110	4 + + : 4 *ad—1 ad—2*
100	4 *ad—1* + : 4 + *ad—2*
80	2 *ad—1* + : 2 + + : 2 *ad—1 ad—2* : 2 + *ad—2*

a. Are the two genes linked?

b. Calculate the distance of each gene from its centromere and, if the genes are linked, the distance between the two genes.

3. Consider two strains in *Neurospora*, *a* + and + *b*. If the genes are linked and *a* is closer to the centromere than *b*, diagram the following and identify the ascus type.

a. no crossover

b. crossover between *a* and the centromere

c. crossover between *b* and the centromere

d. crossover between each gene and the centromere involving two strands

e. crossover between each gene and the centromere involving four strands

f. two crossovers between gene *a* and the centromere, involving four strands

g. two crossovers between gene *b* and the centromere, involving four strands

4. In *Drosophila* there are two linked recessive genes: *y*, which causes yellow body color, and *br*, which causes short bristles. +/*y* is closer to the centromere than +/*br*. If a double heterozygote, in coupling phase, experiences mitotic crossing-over, what are the possible outcomes?

5. Mitomycin C is a drug that increases the frequency of mitotic crossing-over in cultured human cells. How might it be useful in studying rare recessive alleles?

6. Human cells were fused with cells of the mouse. Several different lines with varying numbers of human chromosomes were isolated and tested for the enzyme *X*. The results are tabulated below (+ indicates chromosome or enzyme is present). On which chromosome is gene *X*?

	Human chromosomes										
Cell line	1	2	3	4	5	6	7	8	9	10	Enzyme X
A	+	+	−	−	−	+	+	+	+	+	+
B	+	−	+	+	+	+	+	−	−	+	+
C	−	+	+	+	+	+	−	+	+	+	−
D	−	−	+	−	+	+	−	+	−	−	−
E	−	+	+	−	−	−	+	−	+	−	+

7. Assume that you are studying the effects of what appear to be two different genes but, in reality, the two phenotypic characteristics are caused by a pleiotropic gene. What will be the X^2 result?

Solutions to Self-Test

1. a. PD = 110, NPD = 0. PD >> NPD. NPD/T = 0/18. Therefore, the two genes are linked.

b. Gene *ad* shows only first-division segregation. It is very close to the centromere. *ylo* to centromere distance: 100%(1/2)(18)/128 = 14.1 m.u. *Ad* to *ylo* distance: 100%(1/2)(18)/128 = 14.1 m.u.

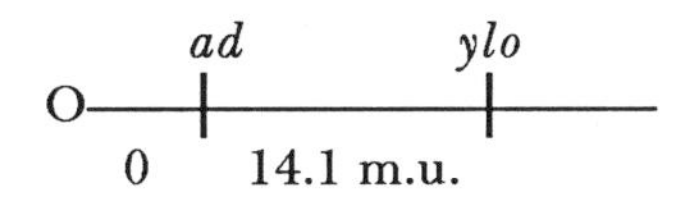

2. a. NPD = 100, PD = 110, T = 80. PD almost equals NPD. NPD/T = 100/80 = 1.25. Therefore, the genes are not linked.

b. *Ad—1* to centromere: (100%)(1/2)(80)/290 = 13.8 m.u. *Ad—2* to centromere: (100%)(1/2)(0)/290 = 0 m.u.

ad—1: O———————|——— 13.8 m.u.

ad—2: O———|——————— 0 m.u.

3. a. 4 *a* + : 4 + *b*, PD

b. 2 *a* + : 2 + *b* : 2 *a* + : 2 + *b*, PD

c. 2 *a* + : 2 *a b* : 2 + + : 2 + *b*, T

d. 2 *a* + : 2 + + : 2 *a b* : 2 + *b*, T

e. 2 + *b* : 2 *a b* : 2 + + : 2 *a* +, T

f. 2 + *b* : 2 + *b* : 2 *a* + : 2 *a* +, PD

g. 4 *a b* : 4 + +, NPD

4. If crossing-over occurs between *y* and the centromere, after recombination both chromosomes would be *y br*/+ +. Depending upon alignment at metaphase, the daughter cells could both be *y br*/+ + (fully wild-type) or one cell could be *y br*/*y br* (single spot that is yellow, with short bristles) and the other could be + +/+ + (fully wild-type).

If crossing-over occurs between the two genes, after recombination one chromosome would be *y br*/*y* + and the other would be + *br*/+ +. Alignment at metaphase could yield the following: *y br*/+ *br* (short bristles) and *y* +/+ + (fully wild-type) or *y br*/+ + (fully wild-type) and *y* +/+ *br* (fully wild-type).

5. By exposing heterozygous cells in culture to mitomycin C, mitotic recombination would result in mitotic segregation 50 percent of the time. If the cells could be cloned, lines homozygous for each allele could be isolated and defined.

6. chromosome 7

7. If you assumed independent assortment, the X^2 will be highly significant. If you assumed some value of linkage, the X^2 will be highly significant if the assumed value was within the range of most mapping problems. If the linkage was assumed to be very tight, a highly significant X^2 would require a very large sample size.

Some Tough Advice On How To Study Genetics

It is my sincere hope you have the habit and the methods of critical analysis that have been presented in both your textbook and this *Companion.* At this point in the text, the material becomes much more descriptive. You now have two main tasks: (1) to memorize the presented material and (2) to integrate the new material with all that has come before. The first task, memorization, should be relatively easy. It requires a skill that you have been perfecting since long before you entered kindergarten.

The second task, integration, is quite difficult for many students. For example, in Chapter 11 you will cover the structure of DNA, in Chapter 14 you will put that DNA into chromosomes, and in Chapter 18 you will study recombination at the molecular level. All of these major topics must be integrated with each other and with what you have learned about recombination at the level of transmission genetics.

If at this point your grasp of transmission genetics is weak, go back to Chapter 2 and learn it. You may be able to memorize the following material without a firm foundation in transmission genetics, you may be able to do very well on tests covering the new material, but, ultimately, that new material will carry little meaning for you unless it can be applied to transmission genetics.

CHAPTER

7

Gene Mutation

Important Terms and Concepts

A gene and its function are not detectable without **variants** to reveal the existence of that gene. Variants arise by two mechanisms: (1) **gene mutation**, which is also known as a **point mutation**, and (2) **chromosome mutation**.

The standard form of a gene is the **wild-type**. Any deviation from the standard form is a **forward mutation**. Any change from the forward mutation back toward the standard form is a **reverse mutation**, a **reversion**, or a **back mutation**.

A **somatic mutation** occurs in nonreproductive cells and leads to a **clone** of cells that differs genetically from the rest of the organism. This results in the organism being classed as a **mosaic** because it has two or more differing cell lines. Somatic mutations can be inherited if the organism undergoes vegetative reproduction.

A **germinal mutation** occurs in reproductive tissue. Germinal mutations can be inherited.

Morphological mutations cause a change in form. **Lethal** mutations result in death. **Conditional** mutations are expressed under the **restrictive** condition and are not expressed under the **permissive** condition. **Biochemical** mutations result in a change in metabolism. Organisms that are nutritionally self-sufficient given a standard set of growth conditions are **prototrophic.** Mutants that require additional supplementation are **auxotrophic.**

The **mutational rate** is the number of mutational events per unit time. The **mutational frequency** is the number of mutant individuals per total number of organisms.

Selective systems aid in the detection of rare mutations.

Evolution occurs through mutations, although most mutations are deleterious.

Mutations can be induced by a number of differing techniques, biochemical and physical. Natural and induced mutations are random in direction, in the gene being affected and in the cell in which they occur.

Be sure that you have thoroughly read the entire chapter before you attempt any of the problems.

Solutions to Problems

1. The petal will now be *Ww*, or blue.

2. Grow cells in the absence of leucine and the presence of an antibiotic that will kill only proliferating cells. Wash out the antibiotic, and then plate the cells on medium containing leucine.

3. Plate the cells on medium lacking proline. Nearly all colonies will come from revertants. The remainder will be second-site suppressors.

4. Because the mutation rate for any given gene is rather low, you will need to screen a large number of organisms exposed to each of the new molecules in order to detect a statistically significant increase in the mutation rate. Assume that a molecule that causes a mutation in microorganisms can cause a mutation in people. Grow large numbers of bacteria under conditions in which they could not grow unless a mutation occurs (an auxotroph in minimal medium or a susceptible bacterium in the presence of a drug), and simultaneously expose them to a new molecule. If the molecule causes mutation, then an increase in survival

should be seen over bacteria not exposed to the molecule. Suspected mutagens can then be tested in animals. This scheme is the basis of the Ames test (Chapter 17).

5. Streak the yeast on minimal medium plus arginine. When colonies appear, replica plate them onto minimal medium. The absence of growth in minimal medium will identify the arginine-requiring mutants.

6. Assume that you are working with twenty specific nutrients that were added to the minimal medium. Group the nutrients, with five to a group. Test each auxotroph with each group. When an auxotroph grows in one of the groups, test the auxotroph separately against each member of the group. A flow sheet would look something like the following:

Test 1	Growth	Test 2	Growth
group A (1 to 5)	–	11	–
group B (6 to 10)	–	12	–
group C (11 to 15)	+	13	+
group D (16 to 20)	–	14	–
		15	–

These results tell you that nutrient 13 was required for growth.

If a mutant cannot be identified with a specific nutrient requirement, then it may be a double or multiple mutant. Alternatively, it could require a nonidentified component of the minimal medium.

7. The first task in this problem is to understand the data. There are nine strains, 1 through 9. Strains 1 and 2 are the parents, 3 through 9 are the progeny. Strains 2, 7, 8, and 9 were each crossed with strains 1, 3, 4, 5 and 6. The results are presented in Table 7-8 in the textbook.

Strain 2 carries a *mei* mutation (it results in 0 when crossed to strains 4 and 6). It must be recessive because some crosses result in full fertility (with 1, 3, and 5). The *mei* mutation was passed to strains 4 and 6 (0 when crossed with strain 2) and 7 and 8 (0 when crossed with either 4 or 6). The *mei* mutation may inhibit pairing.

8. You need to apply the Poisson distribution to answer this problem. Mutants were *not* (zero class) observed on 37 plates out of 100. The formula is $e^{-\mu n}$ = number in zero class/total number. Or $e^{-\mu} \times 10^6 = 37/100$, and $u = -\ln(0.37) \times 10^{-6} = 1/10^{-6}$ cell divisions.

9. There are many ways to carry out this experiment. Using *Drosophila*, you could raise flies with (experimental) and without (control) caffeine added to the diet. You could do the same with mice, rats, cats, dogs, or mammalian cells in culture. Alternatively, you could inject a solution plus or minus caffeine into an organism.

10. To solve this problem, you need to assume that euchromatic chromosomes are genetically active and that heterochromatic chromosomes are genetically inactive. Thus, the female requires ten active chromosomes while the male requires only five. In order to have the normal sex ratio, the data suggest either that heterochromatic chromosomes normally must segregate from euchromatic chromosomes or that paternally derived chromosomes become heterochromatic in males.

When the female parent is irradiated, lethal mutations are induced. The mutations would be dominant in both sexes among the offspring. Thus, there are no progeny.

When the male parent is irradiated, lethal mutations again are induced. They would be dominant in the female offspring but recessive in male offspring, because the chromosomes donated by the male parent to male offspring normally are inactive.

11. Assume that a cell's phenotype is determined by location, not origin. Then cells derived from the epidermis but located in the photosynthetic layers of the leaf should develop chlorophyll. They would be green patches in an albino leaf.

Assume that a cell's phenotype is determined by origin, not location. Because epidermal cells normally do not have chlorophyll, these migrated cells would not be green and would be undetectable in an albino leaf.

12. Stain pollen grains, which are haploid, from a homozygous *Wx* parent. Look for red pollen grains, indicating mutations to *wx*, under a microscope.

13. a. Choose a stage in the life cycle that is easy to irradiate. Either seeds or pollen grains would work easily.

b. Determine the unirradiated survival rate.

c. Determine the irradiated survival rate at several different dosages, then select a dosage that gives a good yield of both offspring and mutations.

d. Observe plants grown from irradiated seeds or pollen grains. You will see only dominant mutations.

e. Self the normal plants (do not lodge) and ignore all the recessive mutations that show in the offspring from this cross.

f. Self the normal progeny from the previous cross. If only normal progeny are obtained, then the parents and the progeny have a good chance of being homozygous. Several more generations of selfing of normal plants may be needed to reach homozygosity.

14. The most straightforward explanation is that a mutation from wild-type to black occurred in the germ line of the male wild-type mouse. Thus, he was a gonadal mosaic of wild-type and black germ cells.

15. a. Mary may have had a chromosome rearrangement (translocation or a large inversion, see Chapter 8) or a dominant mutation induced by the radiation.

b. The chromosome rearrangements can be checked by examining the chromosomes of her descendants who have spontaneous abortions. The dominant mutation would be difficult to confirm practically.

16. An X-linked disorder cannot be passed from father to son. Because the gene for hemophilia must have come from the mother, the nuclear power plant cannot be held responsible.

It is possible that the achondroplastic gene mutation was caused by exposure to radiation.

17. The mutation rate needs to be corrected for achondroplastic parents and put on a "per gamete" basis:

$$\frac{10 - 2}{2 \times 94{,}075} = 4.25 \times 10^{-5} \text{ gametes}$$

You do not have to worry about revertants in this problem because the problem asks for the net mutation frequency to achondroplasia.

18. The commission was looking for induced recessive X-linked lethal mutations, which would show up as a shift in the sex ratio. A shift in the sex ratio is the first indication that a population has sustained lethal genetic damage. Other recessive mutations might have occurred, of course, but they would not be homozygous and therefore would go undetected. All dominant mutations would be immediately visible, unless they were lethal. If they were lethal, there would be lowered fertility and/or an increase in detected abortions, but the sex ratio would not shift as dramatically.

19. There are many specific answers. Any answer would have to include the following:

1. Any increase in radiation level above background will result in an increased mutation rate. This could lead to an increased rate of cancer, leukemia, birth defects, and spontaneous abortions. It is more likely that there will be lowered fertility rather than increased birth defects or detected spontaneous abortions because most mutations might be

expected to interfere either with getting pregnant or with development at very early stages of pregnancy.

2. The increase in malignancy and reproductive difficulties will be directly proportional to the dosage that the population receives.

3. The increase in malignancy and reproductive difficulties will also depend on the genetic constitution of the population receiving radiation.

4. Ultimately, there is not enough known to allow precise predictions for any population exposed to a given amount of radiation above background.

20. From hospital records you can determine the frequency of birth defects, abortions, and malignancies prior to the explosion. You can also determine the sex ratio for newborn children prior to the explosion. These figures can be compared with the frequency of each after the explosion. In addition, you can monitor the population over time for chromosomal abnormalities in the peripheral blood and compare the population exposed to radiation with another, unexposed population that is located within the same general area of the country.

21. **a.** reddish all over

b. reddish all over

c. many small, red spots

d. a few large, red spots

e. like c, but with fewer reddish patches

f. like d, but with fewer reddish patches

g. some large spots and many small spots

Self-Test

1. Recall the replica plating test. Assume that Lederberg and Lederberg observed that one in a thousand colonies were resistant to penicillin when tested by replica plating. When they next tested the colonies that gave rise to the penicillin-resistant colonies, what percentage of the cells in each colony would be expected to be penicillin-resistant if

a. resistance is caused by penicillin?

b. resistance is selected by penicillin?

2. During the 1940s and into the early 1950s in the United States, many shoe stores used a continuous X-ray machine to check shoe fit. Frequently, young children would watch their foot bones wiggle in the machine while their mothers tried on shoes. What effect would be expected from this?

3. How can you distinguish a new mutation from a mutation that had not been expressed in prior generations due to epistasis?

4. A child is born with cleft lip and palate, a disorder known to be caused by some environmental agents and also to have several genetic causes (both dominant and recessive forms exist). The parents state that there is no history of the disorder on either side of the family. Is it likely that the child represents a new mutation?

5. A child is born with neurofibromatosis. This is a dominant genetic disorder with the highest rate of mutation known in humans. The high mutation rate of neurofibromatosis is perhaps explained by the fact that it is also a disorder that has reduced penetrance, has variable expressivity, and is very frequently not diagnosed when present. Before coming to the conclusion that a new mutation has occurred with this particular child, what must you do?

6. Are abnormal phenotypes caused by an environmental agent inherited?

Solutions to Self-Test

1. a. If resistance is caused by exposure to penicillin, then approximately the same frequency of resistance would be observed, 1/1000 cells.

b. If resistance is already present in the colonies and is simply selected by penicillin, then the range of resistant cells can be from very few (only a subpopulation of the colony is resistant) to 100 percent of the cells (the entire colony is resistant).

2. The rapidly growing cells giving rise to toenails would be the cells most likely to experience mutation. This could be expressed as very small nails, deformed nails, absence of nails, and even possibly malignancy.

3. This is very difficult practically in humans, less so in other organisms. In the fruit fly, for instance, if the mutant allele had been blocked from expression, then testcross progeny of sibs of the parents whose progeny first expressed it might also express it. If a new mutation occurred, sibs of the parents whose progeny first expressed it would not be expected to carry the allele.

4. While it is possible that the child represents a new mutation, it is more likely that environmental exposure or genetic inheritance is the cause. Mutation should be used as the explanation of last resort because the rate of mutation is much lower than the frequency of any specific birth defect.

5. The parents and all sibs of the child should have a complete physical examination to rule out neurofibromatosis. Many children with this disorder have relatives who, although not diagnosed, have some symptoms of neurofibromatosis.

6. Those environmental agents that cause mutations would give rise to inherited changes, although the direction of mutation would be random. Those environmental agents that interfere with development without causing mutations (teratogenic agents) would not give rise to inherited changes.

CHAPTER

8

Chromosome Mutation I: Changes in Chromosome Structure

Important Terms and Concepts

Chromosome mutations are changes in the genome involving chromosome parts, whole chromosomes, or whole chromosome sets. They are also known as **chromosome aberrations**. They can be detected either by genetic tests or by viewing the chromosomes under the microscope.

Cytogenetics is the combined study of cells (cytology) and genetics.

Chromosomes can be distinguished by size, centromere position, nucleolar organizers, satellites, and staining patterns.

Specific autosomal chromosomes within a genome are numbered. The larger the chromosome, the smaller the number. If two chromosomes are of the same size, the one with the more centrally positioned centromere has the lower number.

The **primary constriction** is the centromere. The **secondary constriction** is the **nucleolar organizing region**, or NO.

Metacentric chromosomes have the centromere in the middle of the chromosome. **Acrocentric** chromosomes have the centromere off center. **Telocentric** chromosomes have the centromere at the end of the chromosome. **Acentric** chromosomes do not have a centromere. **Dicentric** chromosomes have two centromeres.

A **telomere** is the end of a chromosome. A **satellite** is a small piece of chromosome distal to the nucleolar organizing region.

Heterochromatin is a densely staining region thought to be genetically inert. **Constitutive** heterochromatin does not vary from cell to cell within a species. **Facultative** heterochromatin varies with cell type within a species. **Euchromatin** stains very lightly and is thought to be genetically active.

Banding patterns along the length of chromosomes can occur naturally in some species, such as *Drosophila*, or be induced by various treatments in other species, such as humans.

Endomitosis is a process of chromosome replication not followed by cell division. It leads to an increase in the total number of chromosomes in a cell. In humans, a normal feature of liver and other cells is one round of endomitosis, leading to 92 chromosomes. In *Drosophila* salivary gland cells, endomitosis results in **polytene chromosomes**.

Chromosome rearrangements consist of deletions, duplications, inversions, and translocations.

A **deletion** is the loss of a chromosome segment. A **terminal** deletion results from one chromosome break. An **interstitial** deletion results from two chromosome breaks. The region between the two breaks is lost when chromosome repair occurs. Deletions frequently result in **pseudodominance**, the expression of a recessive gene when present in a single copy. A deletion results in an unpaired loop during synapsis. Homozygous deletions usually are lethal.

A **duplication** consists of more than one copy of a chromosomal segment on one chromosome. Adjacent duplicated segments occur in **tandem sequence** with respect to each other (abcdabcd) or they may occur in **reverse order** with respect to each other (abcddcba). Duplications, like deletions, can disturb the genetic balance of the genome, resulting in abnormal development or function. They also supply additional genetic material capable of evolving new functions. Duplications result in an unpaired loop during synapsis.

Inversions result from two chromosome breaks, with a subsequent "flipping" of the middle segment with respect to the two ends, followed by chromosome repair. **Paracentric** inversions do not involve the centromere. Crossing-over in the inverted region leads to an acentric fragment and a dicentric chromosome which then enters the **breakage-fusion-bridge cycle**, resulting in duplications and deletions. **Pericentric** inversions involve the centromere. Crossing-over in the inverted region leads to duplications and deletions. Inversion heterozygosity results in a reduction of viable recombinant gametes. Inversion heterozygotes have a paired loop during synapsis.

A **translocation** is the movement of a segment of a chromosome to a new location. A **reciprocal** translocation consists of one break in each of two chromosomes followed by an exchange of the acentric fragments. New linkage relations are created by translocations. When nonhomologous chromosomes are involved, pairing at synapsis results in a cross configuration involving four chromosomes in the heterozygote. Synapsis in the homozygote does not result in the cross structure. Translocation heterozygosity leads to greatly reduced fertility because segregation usually results in both duplications and deletions for entire chromosomes.

Position-effect is the alteration in a gene's functioning caused by a change in the gene's location.

Be sure that you have thoroughly read the entire chapter before you attempt any of the problems.

Solutions to Problems

1. **a.** Deletions lead to a shorter chromosome with missing bands, if banded, an unpaired loop during homologous pairing, and the expression of hemizygous recessive alleles.

b. Duplications lead to a longer chromosome with repeated bands, if banded, an unpaired loop during homologous pairing, and there may be disturbed development.

c. Inversions can be detected by banding, show the typical twisted homologous pairing for heterozygotes, and no crossover products are seen for genes within the inversion.

d. Reciprocal translocations can be detected by banding, show the typical cross structure during homologous pairing, lead to new linkage groups, and show altered linkage relationships. The heterozygote has a high rate of unbalanced gamete production.

2. a. In heterozygotes, the products of crossing-over will be nonviable 25 percent of the time. Thus, the RF will be about 27 percent.

b. The homozygote will have no trouble with crossing-over, so the RF will be about 36 percent.

3. P *A— B— C— D— E— F—* × *aa bb cc dd ee ff*

F_1 1/2 *A— B— C— D— E— F—*

1/2 *A— B— C— dd ee F—*

Remember that these genes are linked. Because all progeny flies are *A– B– C– F–*, the wild-type must have been homozygous for these genes. This means that half the progeny received *D E* and half received *d e* from the wild-type parent. No recombinants were seen. The best explanation is that the wild-type fly was heterozygous *D E/d e* and that an inversion spanned these two genes.

4. a. paracentric inversion

b. deletion

c. pericentric inversion

d. duplication

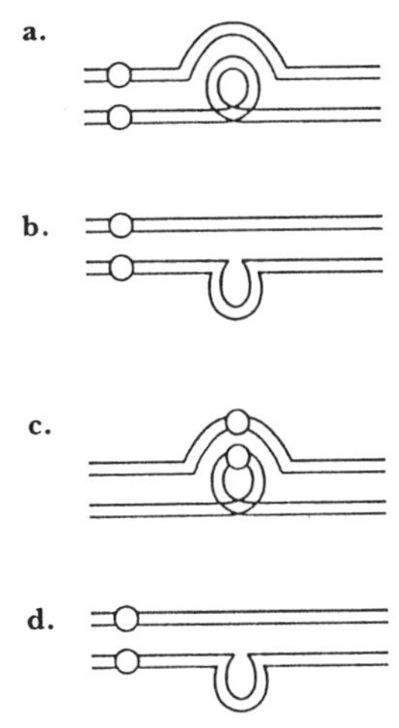

5. From each of the statements concerning the rare cells, you should be able to draw the following conclusions:

Statement	*Conclusion*
require leucine	*leu*$^+$ lost
do not mate	one mating type lost
will not grow at 37°C	*un*$^+$ lost
cross only with *a* type	*a* lost
only nucleus 1 recovered	deletion occurred in nucleus 2

Because *ad—3A*$^+$ and *nic*$^+$ function are required by the heterokaryon, these genes must have been retained. Therefore, the most reasonable explanation is that a deletion occurred in the left arm of chromosome 2 and that *leu*$^+$, mating type *a*, and *un*$^+$ were lost.

6. The single waltzing female that arose from a cross between waltzers and normals is expressing a recessive gene when it is present in one copy. That means that the normal allele must have been deleted.

When she was mated to a waltzing male, all the progeny were waltzers. This further supports a deletion of the normal allele. Had it been present, one-half of the progeny would be normal.

When she was mated to a normal homozygous male, all the progeny were normal. This indicates that her waltzing was due to a recessive allele and does not represent a dominant variant for waltzing. Half of these offspring are hemizygous normal and half are heterozygous normal, if a deletion occurred.

When some of these offspring are mated, the progeny are normal. The offspring are w^+w (normal) or w^+w^* (normal, with deletion). Without a deletion in the normal female, one-fourth of their progeny would be waltzers. With a deletion, progeny homozygous for a deletion would be expected to die. Progeny that are ww^* would be expected to be waltzers. These waltzers were not observed. Because this is the only discrepancy, it must be assumed that not enough crosses were performed.

The cytological data also support a deletion because the abnormal waltzing female and her abnormal progeny had one member of a chromosome pair that was abnormally short.

7. The colonies that would not revert most likely had a deletion within the *ad—3B* gene. If the gene had not been deleted, at least some reversion would have been seen. Because the gene was deleted, however, there was nothing there for the mutagens to work on. They could grow with adenine supplementation only.

8. Compare deletions 1 and 2: allele *b* is more to the left than alleles *a* and *c*. The order is *b* (*a c*), where the parentheses indicate that the order is unknown.

Compare deletions 2 and 3: allele *e* is more to the right than (*a c*). The order is *b* (*a c*) *e*.

Compare deletions 3 and 4: allele *a* is more to the left than *c* and *e*, and *d* is more to the right than *e*. The order is *b a c e d*.

Compare deletions 4 and 5: allele *f* is more to the right than *d*. The order is *b a c e d f*.

Allele	Band
b	1
a	2
c	3
e	4
d	5
f	6

9. When a deletion is crossed with a recessive point mutation in the same gene, the recessive point mutation is expressed. When the deletion and the point mutation are in different genes, wild-type is observed. This is exactly like the results seen in a cross between two allelic variants (non-wild-type progeny) and a cross between organisms that have defects in two different genes (wild-type progeny).

Mutant	Defect
1	deletion of at least part of genes *h* and *i*
2	deletion of at least part of genes *k* and *l*
3	deletion of at least part of gene *m*
4	deletion of at least part of genes *k*, *l*, and *m*
5	a deletion not within the *h* through *m* genes or a recessive point mutation

10. a. Eighteen map units (m.u.) were either deleted or inverted. A large inversion would result in semisterility, whereas a large deletion would most likely be lethal. Thus, an inversion is more likely.

b.

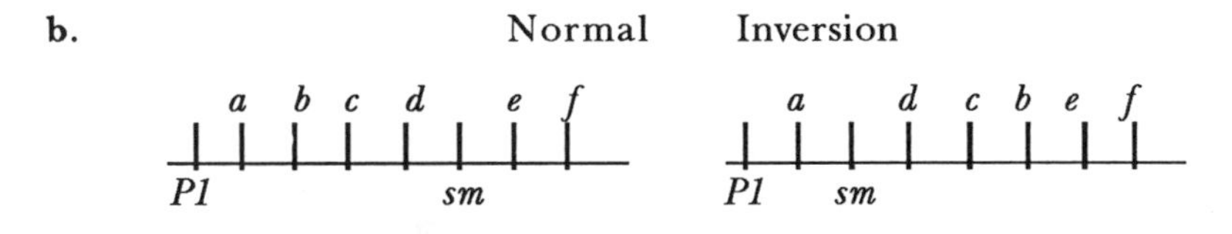

c. The semisterility is the result of crossing-over in the inverted region. All products of crossing-over would have both duplications and deletions.

11. The testcross is $P\ B\ Q/p\ B\ q \times p\ b\ q/p\ b\ q$.

a. To obtain normal eye shape, the bar allele must be deleted. Actually, bar eye is the result of a tandem duplication of a normal allele. Thus, to delete the extra copy of the gene, synapsis must occur out of register between the two chromosomes:

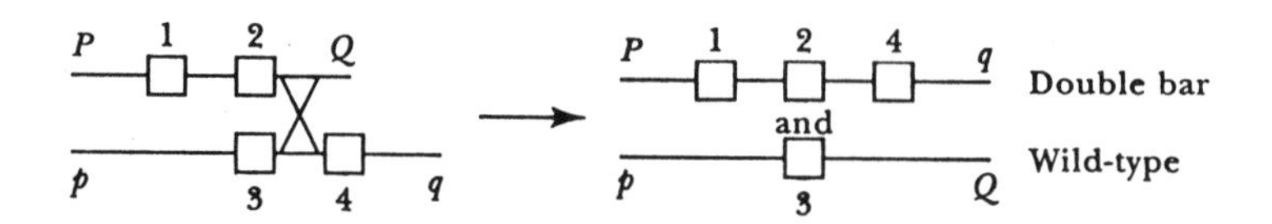

b. In the diagram above, the flanking markers are $P\ q$ and $p\ Q$ after crossing-over. If gene 1 had paired with gene 4, then the wild-type phenotype would have been associated with $P\ q$, and the double bar would have been with $p\ Q$.

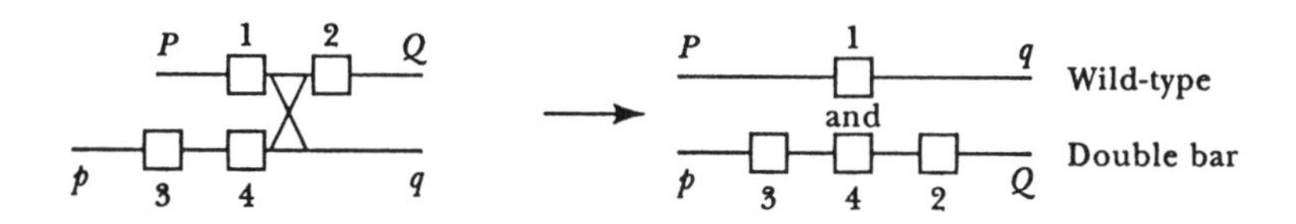

12. **a.** If the chromosome paired with itself, as diagramed below, somatic crossing over would yield the observed result.

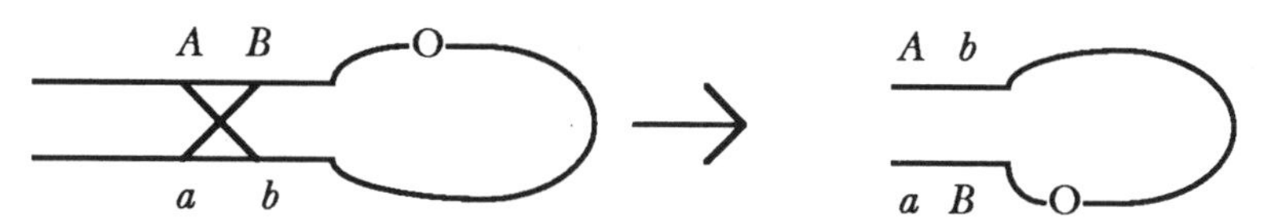

b. Because the culture is haploid, only one copy of the chromosome is present. This eliminates the possibility of crossing over between homologous chromosomes. Crossing over could occur between B/b and the centromere of the chromosome diagramed above, giving the result shown below.

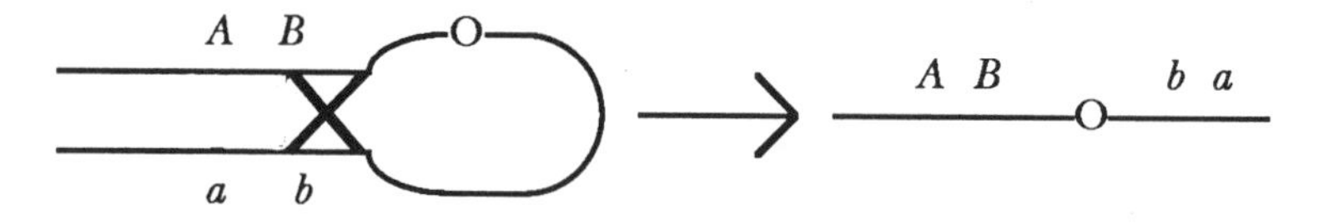

Unless nondisjunction first occurs, so that crossing-over could take place between homologous chromosomes, no other crossover products would be seen. If nondisjunction occurred, pairing would be as follows:

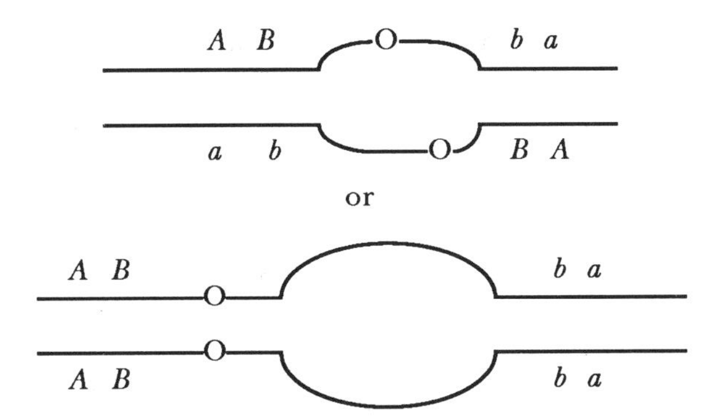

Single and double crossovers could give many different results, depending on where they occur.

13. Classical dominance states that one copy of the dominant allele is sufficient for the dominant phenotype. In this case, the female progeny have one copy of the dominant allele, yet they do not have the dominant phenotype. To explain these results, it must be assumed that both the dominant and the variant alleles have a product and that phenotype is the result of the ratio of one product to the other.

a. The female parents are $v^+ v^- / v^+ v^-$. The ratio of the two alleles is 1:1, which results in the wild-type phenotype. The male parents are v^-/Y. No dominant alleles are present and the males are vermilion.

b. The male progeny are $v^+ v^-$/Y, and the ratio of the two alleles is 1:1, yielding a wild-type phenotype identical to that of their female parents. The female progeny are $v^+ v^- / v^-$. The ratio of dominant alleles to vermilion is 1:2. Thus, they have the vermilion phenotype.

14. **a.** Single crossovers lead to tetratypes. Also, only one-half of the tetratype asci are crossover products, so the 10 m.u. must be multiplied by 2 to yield the right frequencies.

+ +, + +, + +, + +, *un3 ad—3*, *un3 ad-3*, *un3 ad—3*, *un3 ad—3* 80%

+ +, + +, + *ad-3*, + *ad—3*, *un3* +, *un3* +, *un3 ad-3*, *un3 ad—3* 10%

+ +, + +, + *ad—3*, + *ad—3*, *un3 ad—3*, *un3 ad—3*, *un3* +, *un3* + 10%

b. The aborted spores result from an inversion in the wild-type. Crossing-over led to nonviable spores because they were unbalanced. This could be tested by selecting *un3 ad—3* double mutants from the wild-type and then crossing them with the + + inverted strain. The *un3* to *ad—3* distance should be altered.

15. **a.** To construct the maps, look at three genes at a time. For example, the Okanagan sequence for *A*, *B*, and *C* is

A to *B*: 12 m.u.

The only possible sequence is *A B C.*

		20		3		12		2		15	
Okanagan sequence:	D	____	E	____	A	____	B	____	C	____	F

		15		4		3		12		18	
Spain sequence:	F	____	C	____	E	____	A	____	B	____	D

b. Diagram the heterozygote during homologous pairing. The inversion cannot be placed exactly. The following assumes that crossing-over can occur only between *C* and *F.*

	A	*B*	*C*	*D*	*E*	*F*
A	0	0	0	0	0	15
B		0	0	0	0	15
C			0	0	0	15
D				0	0	15
E					0	0
F						0

16. a. The aberrant plant is semisterile, which suggests an inversion. Since the *d—f* and *y—p* frequencies of recombination in the aberrant plant are normal, the inversion must involve *b* through *x.*

b. To obtain recombinant progeny when an inversion is involved, either a double crossover occurred within the inverted region or single crossovers occurred between *f* and the point of inversion, which occurred someplace between *f* and *b.*

17. a. $0.1 = 1/2(1 - e^{-m})$, or $m = -\ln(1 - 0.2) = 0.22$

b. The better general formula is $f(i) = e^{-m}m^i/i!$

1. $f(0) = e^{-0.22}(0.22)^0/0! = 0.80$
2. $f(1) = e^{-0.22}(0.22)/1 = 0.176$
3. $f(2) = e^{-0.22}(0.22)^2/2 = 0.0042$

c. 1. Because no crossovers occur, the ratio of dark to light will be 8:0.

2. One crossover will result in a 4:4 ratio since a crossover within the inversion loop will result in all recombinant products being unbalanced.

3. Two crossovers will result in three different ratios, in a 1:2:1 overall ratio: 1 (8:0) : 2 (4:4) : 1 (0:8).

Number of crossovers	Ratios among asci 8:0	4:4	8:0
0 (0.8)	0.8	0	0
1 (0.176)	0	0.176	0
2 (0.0042)	0.0011	0.0021	0.0011
Total	0.8011	0.1781	0.0011

18. The F_1 females are *y cv v f+ car*/+ + + + + +. These are crossed with *y cv v f B car*/Y males.

Class 1: parental

Class 2: parental

Class 3: DCO *y—cv* and *B—car*

Class 4: reciprocal of class 3

Class 5: DCO *cv—v* and *v—f*

Class 6: reciprocal of class 5

Class 7: DCO *cv—v* and *f—car*

Class 8: reciprocal of class 7

Class 9: DCO *v—cv* and *v—f*

Class 10: reciprocal of class 9

Class 11: This class is identical to the male parent's X chromosome and could not have come from the female parent. Thus, the male sperm must have donated it to the offspring. In *Drosophila,* sex is determined by the ratio of X chromosomes to the number of sets of autosomes. The ratio in males is 1X:2A, where A stands for the autosomes contributed by one parent (the ratio in females is 2X:2A). Thus, this class of males must have arisen from the union of an X-bearing sperm with an egg that was the product of nondisjunction for X and contained only autosomes. This class should have only one sex chromosome, which could be checked cytologically.

19. The inversion results in no viable crossover products from heterozygous females. When *Cu pr*/ *Cu pr* females are crossed with irradiated wild-type males, all female progeny will be heterozygous for the inversion and for any recessive lethal mutation induced by irradiation. They will have curled wings and wild-type eyes (*Cu pr*/+ + ?). Each female will carry a different mutation (?), if any were induced. Cross the females individually with a homozygous *Cu pr* male to generate groups of flies with the same mutation. Then, cross the normal-eyed progeny among themselves (*Cu pr*/+ + ?). This results in

1/4 + + ?/+ + ?	normal wings and eyes
1/2 *Cu pr*/+ +	curled wings and normal eyes
1/4 *Cu pr*/ *Cu pr*	curled wings and purple eyes

If a lethal mutation had been induced, the class with normal wings and eyes will be missing.

20.

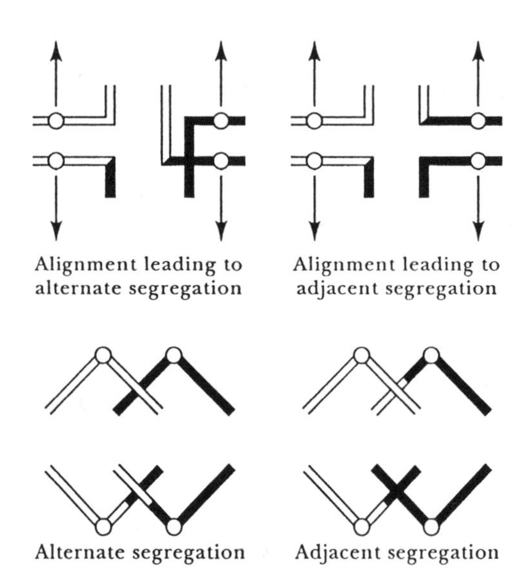

21. If the *a* and *b* genes are on separate chromosomes, independent assortment should occur giving equal frequencies of *a b*, + +, *a* + and + *b*. This was not observed; instead, the two genes are behaving as if they were linked, with 10 m.u. between them. This is indicative of a reciprocal translocation in one of the parents, most likely the wild-type from nature.

At meiosis prior to progeny formation, the chromosomes would look like the following (centromere not included because its position has no effect on the results).

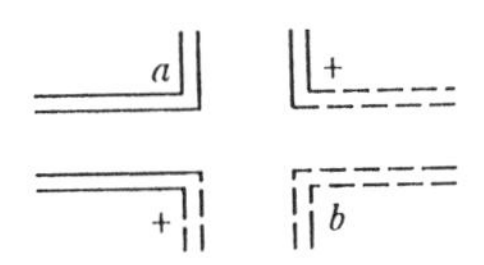

Only alternate segregation avoids duplications and deletions for many genes. Therefore, the majority of the progeny would be parental *a b* and +

+. The *a* + and + *b* progeny would result from crossing-over between either gene and the breakpoint locus.

22. Notice that the males have the male parent phenotype while the females have the female parent phenotype. This suggests a translocation of the *Cy* chromosome to the Y chromosome.

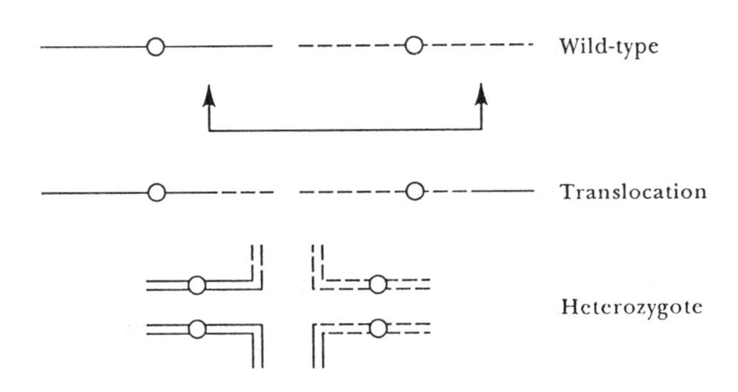

23. a.

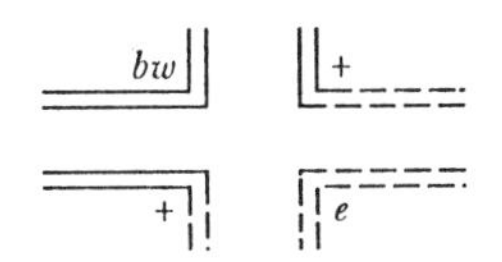

b. The surviving offspring would result from alternate segregation and would be either *bw e* (brown eye, ebony body) or + + (wild-type), in a 1:1 ratio.

24. The size of the insertional translocation will determine whether the translocated region or the rest of the region will dominate homologous pairing. Below, it is assumed that the insertion is quite small in relation to the rest of both chromosomes.

On p. 148, the following asci types will be seen:

Type A: 8:0

translocated

translocated

normal

normal

Type B: 4:4

duplication

duplication

deletion

deletion

Type C: 6:2

translocated

deletion

duplication

normal

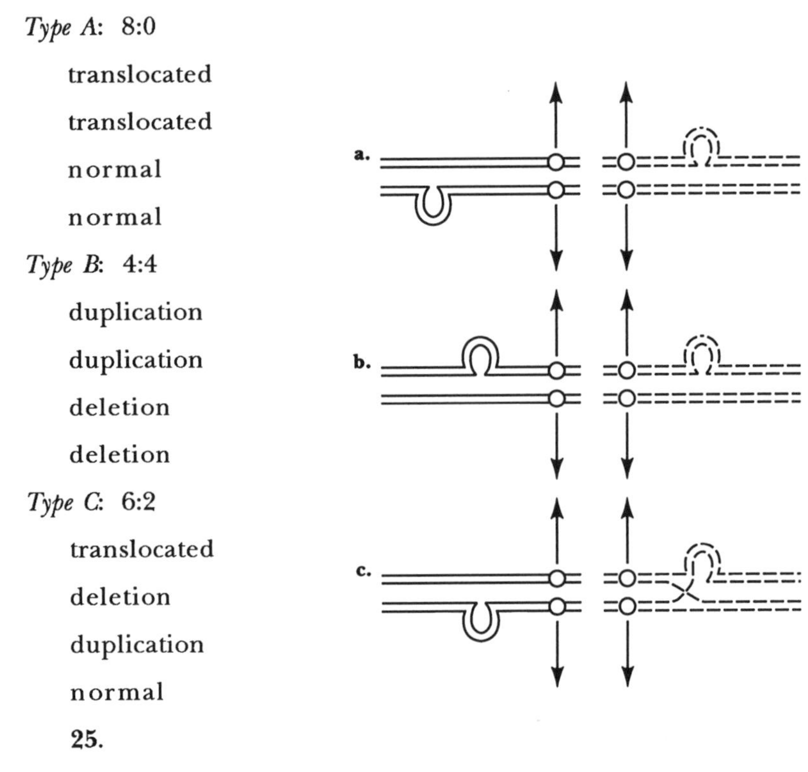

25.

Cross 1: Independent assortment of 2 genes occurred.

Cross 2: Two genes are linked at 1 m.u. distance. Therefore, a reciprocal translocation took place and both genes were very close to the breakpoint. The black spores resulted from alternate segregation, the white from adjacent segregation.

Cross 3: Half the spores were normal, and non-translocated, and half contained both translocated chromosomes.

26. The original plant was homozygous for a reciprocal translocation that brought genes *P/p* and *S/s* very close together. Because of the close linkage, a ratio suggesting a monohybrid, instead of a dihybrid, cross was observed, both with selfing and with a testcross. All gametes are fertile because of homozygosity.

original plant: *P S/p s*

tester: *p s/p s*

F_1 progeny: heterozygous for the translocation:

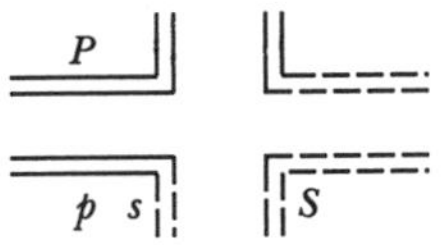

P

p s S

The easiest way to test this is to look at the chromosomes of heterozygotes during meiosis I.

27. The breakpoint can be treated as a gene with two "alleles," one for normal fertility and one for semisterility. The problem thus becomes a two-point cross.

parentals	764 semisterile *Pr*
	727 normal *pr*
recombinants	145 semisterile *pr*
	186 normal *Pr*
	1822

$$100\%(145 + 186)/1822 = 18.17 \text{ m.u.}.$$

28.

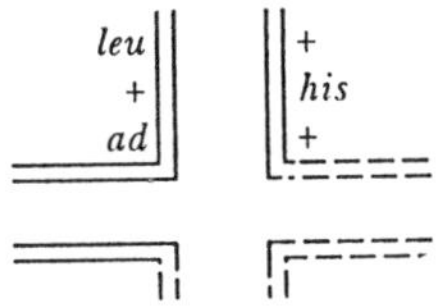

leu + ad + his +

Because the short arm carries no essential genes, adjacent—1 segregation will yield progeny that are viable. Select for *leu*$^+$, *his*$^+$ and *ad*$^+$ by omitting those components from a minimal medium.

29. a. Breaks in different regions of 17R result in deletion of all genes from the breakpoint.

b. Because there is only 17R from humans, all the human genes expressed must be on 17R.

Notice that only gene *c* is expressed by itself. This means that gene *c* is closest to the mouse material. Next notice that if *c* and one other gene are expressed, that other gene is always *b*. This puts *b* closer to mouse material than *a*. The gene order is mouse – *c* – *b* – *a*.

The probability of a break between two genes is a function of the distance between them. Of the 200 lines tested, 48 expressed no human activity. Thus, the *c* gene is no more than (100%)(48)/200 = 24 relative m.u. from the mouse material. A break between c and *b* (express *c* only) occurred in 12 lines, placing these genes (100%)(12)/200 = 6 relative m.u.

apart. A break between *b* and *a* (express *c* and *b*) occurred in 80 lines, placing these genes (100%)(80)/200 = 40 relative m.u. apart. Finally, 60 clones expressed all three genes, placing gene *a* (100%)(60)/200 = 30 relative m.u. from the end of the chromosome.

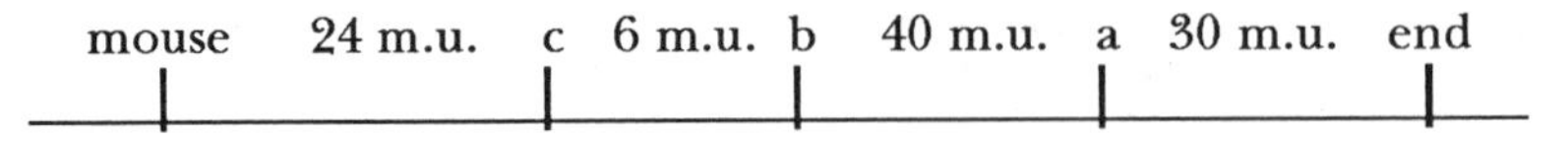

c. The dye could be used to correlate band presence with gene presence.

30. a.

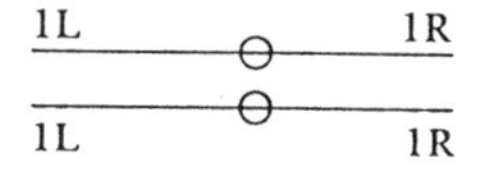

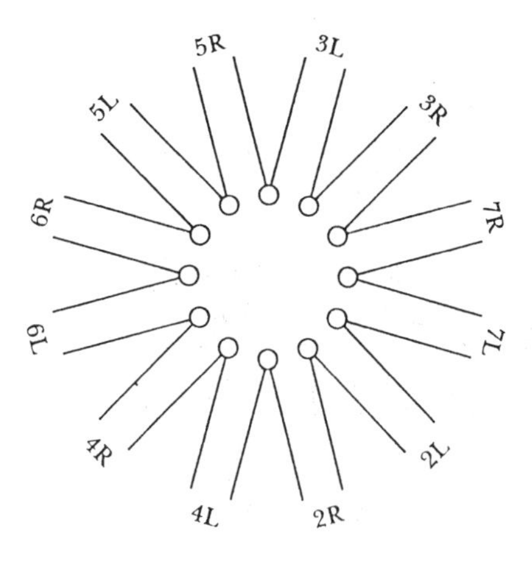

b. Alternate segregation yields balanced gametes.

c. Any mechanism that you can think of is a good explanation. This plant actually uses a system of balanced lethals to maintain heterozygosity.

d. Probably the best answer is that heterozygosity is promoted and heterozygosity leads to hybrid vigor.

31. Species B is probably the "parent" species. A paracentric inversion in this species would give rise to species D. Species E could then occur by a translocation of *z x y* to *k l m*. Next, species A could result from a translocation of *a b c* to *d e f*. Finally, species C could result from a pericentric inversion of *b c d e*.

32. The assumptions are that half of the gametes from a single heterozygous translocation are nonviable and that the two parents have the same chromosomes involved in translocations.

The progeny of crosses between parents with translocations is referring to the F_1 progeny, in which the parental generation was heterozygous. Designate the parents as follows:

A: (T1 T2 N1 N2)

B: (T1 T2 N1 N2)

The gametes will be

A:		B:	
	1/4 (T1 T2)		1/4 (T1 T2)
	1/4 (N1 N2)		1/4 (N1 N2)
	*1/4 (T1 N2)		*1/4 (T1 N2)
	*1/4 (T2 N1)		*1/4 (T2 N1)

where * equals an unbalanced gamete.

Fertilization between balanced gametes will occur 4(1/4)(1/4), or 4/16 of the time. An additional 2/16 of the fertilizations will lead to a balanced fetus even though both gametes were unbalanced; example: (T2 N1) × (T1 N2). Therefore, 6/16 of the progeny are viable and 10/16 are nonviable.

Self-Test

1. A new mutant appears in mice that causes a rough coat. Chromosome analysis reveals that one number 3 chromosome is slightly shorter than its homolog and that two interstitial chromosome bands are missing. What is the best interpretation of the new mutant?

2. When the mutant in Problem 1 above is crossed with a wild-type mouse, all the progeny are wild-type. A cross of F_1 progeny with the mutant parent results in approximately 50 percent stillborn offspring. Explain these results.

3. A dicentric chromosome has the following gene sequence:

a b c   *d e f g h*   *i j k*

————0——————0————

If, at anaphase, a break occurred between genes *f* and *g*, what would be the chromosome composition of each daughter cell after the S phase?

4. In the following inversion heterozygote, diagram the consequences of crossing-over between genes *D* and *E*.

```
A     B     C     D     E     F     G     H
----------------------------O--------------
-----------------------O-------------------
a     b     g     f     e     d     c     h
```

5. Discuss the role that inversions might have played in evolution.

6. In species X, it is known that gene *A* is 10 m.u. from gene *B*. When a heterozygote in repulsion was testcrossed, the following was observed:

490	*Aa bb*, normal fertility
500	*aa Bb*, semisterility
4	*aa bb*, normal fertility
6	*Aa Bb*, semisterility

What can be concluded about genes *A* and *B*?

7. A man had two different wives and, altogether, produced six early miscarriages and two stillborn children with multiple abnormalities. What is the best conclusion regarding the man?

8. The chromosomes of the man in Problem 7 were examined. It was found that he had the following for chromosomes 8 and 9. Diagram synapsis. The dot indicates the centromere.

Chromosome 8: *A.BCDEFG321* and *a.edcbfghi*

Chromosome 9: *1234.56* and *ih4.56*

Solutions to Self-Test

1. A deletion.

2. The stillborn offspring were homozygous for the deletion. The deletion is a recessive lethal.

3. Prior to synthesis the daughter cells would be *abc.def* and *gh.ijk*, where the dot indicates the centromere. The *f* and *g* ends would behave as if they were "sticky" during S, resulting in the following dicentric chromosomes:

abc.deffed.abc duplication of *abcdef*, deletion of *ghijk kji.hggh.ijk* duplication of *ghijk*, deletion of *abcdef*

4. The two strands not involved in crossing-over would give rise to parental chromosomes. The recombinant strands would be

ABCDe.fgba duplication of *ab*, deletion of *h*

HGF.Edch duplication of *h*, deletion of *ab*

5. Inversion heterozygotes cannot yield viable recombinant products for the region that is inverted. They give rise only to parental gametes. Homozygotes for the normal or the inverted sequence have no difficulty in producing recombinant gemetes. Therefore, an inversion acts as a barrier to subpopulations within a species. Over time, two subpopulations may diverge sufficiently to generate two separate species.

6. More than likely, one of the genes has been translocated to a nonhomologous chromosome. The *a B* chromosome was the one involved.

7. The man had either a large inversion or a translocation.

8. The man had both a translocation and an inversion, as diagramed below.

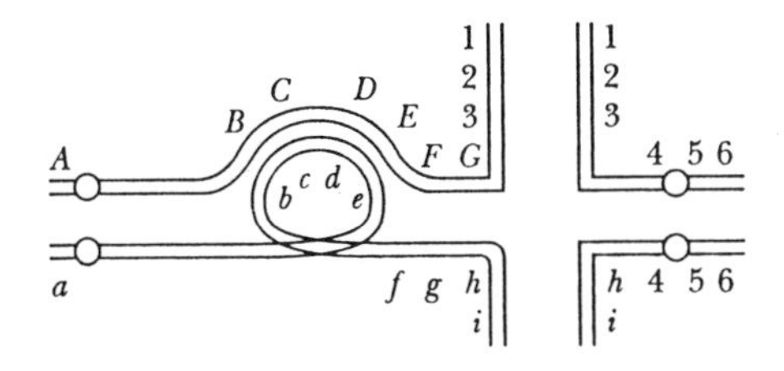

CHAPTER

9

Chromosome Mutation II: Changes in Chromosome Number

Important Terms and Concepts

The **monoploid number** is the number of chromosomes in the basic set of chromosomes. It usually equals the number of chromosomes in a gamete. **Euploid** organisms have integer multiples of the monoploid number. A **diploid** organism has twice the monoploid number of chromosomes. A **haploid** organism has one monoploid number of chromosomes. A **polyploid** organism has more than twice the monoploid number of chromosomes. Examples of polyploids are **triploid, tetraploid, pentaploid, and hexaploid.**

An **aneuploid** has more or less chromosomes than an integer multiple of the monoploid number. Addition of one or more chromosomes produces a **hyperdiploid**. Loss of one or more chromosomes produces a **hypodiploid**. A **monosomic** lacks one chromosome. A **nullisomic** lacks both homologous chromosomes. A **trisomic** has an extra chromosome in a diploid. A *disomic* has an extra chromosome in a haploid.

Colchicine and its less toxic derivative colcemid disrupt the mitotic spindle and block mitosis. They produce polyploids.

Polyploids that have an even-integer multiple of the monoploid number are fertile because each chromosome can pair during meiosis. Those that have an odd-integer multiple of the monoploid number are infertile because one chromosome from each of the multiple homologous chromosomes cannot pair during meiosis.

An **autopolyploid** is composed of multiple sets of chromosomes from one species. The diploid derivative is fertile. An **allopolyploid** is composed of multiple sets of chromosomes from different species. An **amphidiploid** is an allopolyploid from two species. The diploid intermediate is not fertile.

Plants tolerate both polyploidy and aneuploidy much more easily than higher animals.

Be sure that you have thoroughly read the entire chapter before you attempt any of the problems.

Solutions to Problems

1.

Klinefelter's	XXY male
Down's	trisomy 21
Turner's	XO female

2. 1. Develop tetraploid lines with colchicine. Cross an *AAAA* with an *A'A'A'A'*

 2. Cross *AA* with *A'A'*, and then double the chromosomes with colchicine treatment.

3. **a.** 3, 3, 3, 3, 33, 33, 0, 0

 b. 7, 7, 7, 7, 8, 8, 6, 6

4. **a.** If a 6*x* were crossed with a 4*x*, the result would be 5*x*.

 b. Cross *AA* with *aaaa* to obtain *Aaa*.

c. The easiest way is to expose the Aa^* plant cells to colchicine for one cell division. This will result in a doubling of chromosomes to yield AAa^*a^*.

d. Cross 6*x* (*aaaaaa*) with 2*x* (*Aa*) to obtain *Aaaa.*

e. Obtain haploid cells from a plant and obtain resistant colonies by exposing them to the herbicide. Then expose the resistant colonies to colchicine to obtain diploids.

5. The polar nuclei can have the following combinations: *AA BB, AA bb, aa BB,* or *aa bb.* Each can be fertilized by one of the following male gametes: *A B, A b, a B* or *a b.*

6. b.

7. To solve this problem, first recognize that *B* can pair with *B* one-third of the time (leaving *b* to pair with *b*) and that *B* can pair with *b* two-thirds of the time. If *B* pairs with *B*, the result is *Bb, Bb, Bb, Bb,* occurring one-third of the time. If *B* pairs with *b*, the resulting tetrad can be of two equally frequent types, depending on how the pairs align with respect to each other. One type is *BB, BB, bb, bb.* The second is *Bb, Bb, Bb, Bb.*

8. When there is no pairing of chromosomes from the same parent, the tetraploid *AAaa* will produce 1/4 *aa* gametes. The *aaaa* progeny will occur (14)(1/4) = 1/16, and the *A*___ progeny will ooccur 15/16. The phenotypic ratio would be 15:1. With pairing from the same parent only, the tetraploid *BBbb* would produce all *Bb* gametes. The progeny would be 100% *BBbb* and have a phenotypic ratio of 16:0.

9. Large univalents are seen whenever *G. thurberi* is involved in a cross. Small univalents are seen whenever *G. herbaceum* is involved in a cross. Bivalents, large and small, are seen whenever *G. hirsutum* is involved in a cross. The data suggest that *G. thurberi* has 26 small chromosomes, *G. herbaceum* has 26 large chromosomes, and that *G. hirsutum* has 26 large and 26 small chromosomes. *G. hirsutum* is a polyploid derivative of a cross between the two Old World species. This could easily be checked by looking at the chromosomes.

10. a. To do this problem you must first recognize that each allele can pair with any other allele within a gene. For a moment, pretend that you can distinguish all four alleles and, for simplicity's sake, number them 1 (*F*), 2 (*F*), 3 (*f*), and 4 (*f*). The combinations now become 1—2, 1—3, 1—4, 2—3, 2—4 and 3—4. In other words, for each gene there are six combinations. Changing the numbers into letters, the gametes for *F/f* would be: 1/6 *FF,* 4/6 *Ff* and 1/6 *ff.*

When two genes are considered, the gametes are

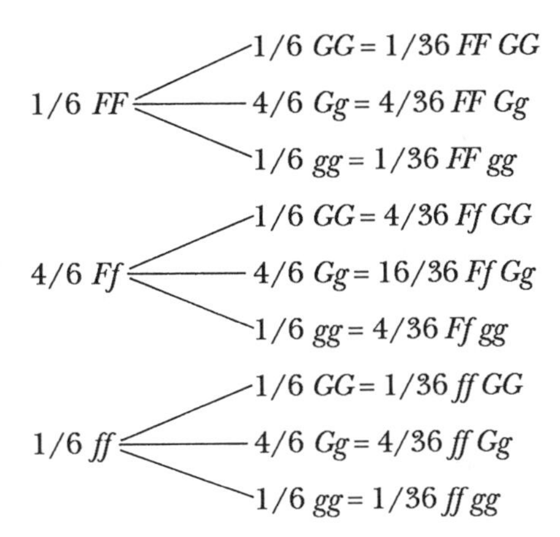

b. The cross is *FFff GGgg* × *FFff GGgg*. For *FFFf GGgg*, consider each gene separately. The combination *FFFf* can be achieved in two ways.

$$\begin{aligned} p(FFFf) &= [p(FF) \times p(Ff)] + [p(Ff) \times p(FF)] \\ &= (1/6 \times 4/6) + (4/6 \times 1/6) \\ &= 4/36 + 4/36 = 8/36 = 2/9 \end{aligned}$$

The combination *GGgg* can be achieved in three ways.

$$\begin{aligned} p(GGgg) &= [p(GG) \times p(gg)] + [p(gg) \times p(GG)] + [p(Gg) \times p(Gg)] \\ &= (1/6 \times 1/6) + (1/6 \times 1/6) + (4/6 \times 4/6) \\ &= 1/36 + 1/36 + 16/36 = 1/2 \end{aligned}$$

Therefore, the $p(FFFf\ GGgg) = 2/9 \times 1/2 = 1/9$

The $p(ffff\ gggg) = p(ffff) \times p(gggg) = 1/6 \times 1/6 \times 1/6 \times 1/6 = 1/1296$

11. e. Only achondroplasia is a gene disorder rather than a disorder characterized by an abnormal chromosome number.

12. One of the parents of the woman with Turner's syndrome (XO) must have been a carrier for color blindness, an X-linked recessive disorder. Because her father has normal vision, she could not have obtained her sole X from him. Therefore, nondisjunction occurred in her father. The sperm lacking an X chromosome fertilized the egg carrying the color blind allele. The nondisjunction event could have occurred during either meiotic division.

If the color blind patient had Klinefelter's syndrome (XXY), then both X's must carry the allele for color blindness. Therefore, nondisjunction had to occur in the mother. Remember that during meiosis I, given no crossover between the gene and the centromere, allelic alternatives separate from each other. During meiosis II, identical alleles on sister chromatids separate. Therefore, the nondisjunctive event had to occur during meiosis II because both alleles are identical.

13. a. If most individuals were female, this suggests that the normal allele has been lost (by nondisjunction or deletion) or is nonfunctional (a process called X-inactivation) in the color blind eye.

b. If most of the individuals were male, this suggests that the male might have two or more cell lines (he is a mosaic, X^{normal} – Y, X^{cb} – Y) or that he has two X chromosomes (he has Klinefelter's syndrome) and that the same processes as in females could be occurring.

14. Assume that simple (nontranslocation) Down's syndrome is the case in both parents; that is, they have 47 chromosomes and are trisomy 21. Each parent has an equal probability of producing gametes with one- and two-chromosome 21. The zygotes can have

p(two 21) = 1/4 diploid (normal)

p(three 21) = 1/2 trisomy (Down's syndrome)

p(four 21) = 1/4 tetrasomy (lethal)

15. If the fluorescent spot indicates a Y chromosome, then two spots are indicative of nondisjunction. Presumably, exposure to dibromochloropropane increases the rate of nondisjunction. This could be tested in several ways. The most straight forward would be to exposes male animals to the chemical, look for an increase in double-spotted sperm over those not exposed, and also do testicular squashes to observe the rate of nondisjunction with and without exposure. Alternatively, specific crosses could be set up that would reveal nondisjunction upon exposure to the chemical. Using X-linkage in fruit flies, white-eyed females exposed to the chemical should have a higher rate of nondisjunction than nonexposed females. When crossed to red-eyed males, nondisjunction of the X chromosome would result in white-eyed females and red-eyed males.

16. Individuals with Down's syndrome resulted from nondisjunction. Leukemia is characterized, in part, by an increase in nondisjunction. It may be that an inherited tendency toward nondisjunction results in both events. Alternatively, the initial unbalancing of the genome with Down's syndrome may lead to progressive unbalancing, ultimately resulting in the leukemic state. There are many other hypotheses that could be proposed.

17. Assuming that other aneuploidies are incompatible with survival to birth, they would best be studied in spontaneously aborted material. It is unlikely that other aneuploidies occur at a lower rate than those seen at birth; it is highly likely that the rate of nondisjunction is approximately the same for all chromosome pairs.

18. One possibility is that the mean age of mothers at birth dropped significantly. Because the older mother is at higher risk for nondisjunction, this would result in the observation. Hospital records could be used to check the age of mothers at birth between 1952 and 1972, as compared with a 20-year period prior to 1952. Another possibility is an increase of amniocentesis amongst older mothers followed by induced abortion of trisomy 21 fetuses. Because pregnant women 35 and older routinely undergo amniocentesis, the rate for this population may have fallen while the rate for the younger population remained unchanged. This also could be checked through hospital records.

19. **a.** Loss of one X in the developing fetus after the two-celled stage.

b. Nondisjunction leading to Klinefelter's syndrome (XXY) followed by a nondisjunctive event in one cell for the Y chromosome after the two-celled stage, leading to XX and XXYY.

c. Nondisjunction for X at the one-celled stage.

d. Either fused XX and XY zygotes or fertilization of an egg and polar body by one sperm bearing an X and another bearing a Y, followed by fusion.

e. Nondisjunction of X at the two-celled stage or later.

20. Remember that nondisjunction at meiosis I leads to the retention of both chromosomes in one cell, while at meiosis II it leads to the retention of both sister chromatids in one cell.

1. Trisomy 21; nondisjunction at meiosis II in the female.

2. Trisomy 21; nondisjunction at meiosis I in the female.

3. Normal; normal meiosis in both parents.

4. Trisomy 21; nondisjunction at meiosis II in the male.

5. Normal; nondisjunction in the female (meiosis I) and the male (either meiotic division).

6. XYY nondisjunction in the male at meiosis II for the sex chromosomes.

7. Monosomy 21-trisomy 21 in a male zygote; mitotic nondisjunction for the 21^b chromosome occurred fairly early in development.

8. Sexual mosaic: fused XX and XY zygotes or, as in Problem **19d,** fused fertilized egg and fertilized polar body.

21. The generalized cross is *AAA* × *aa,* from which *AAa* progeny were selected. These progeny were crossed with *aa* individuals, yielding the results in the table in the textbook. Assume for a moment that each allele can be distinguished from the other, and let 1 = *A*, 2 = *A* and 3 = *a*. The gametic combinations possible are

1 – 2 (*AA*) and 3 (*a*)

1 – 3 (*Aa*) and 2 (*A*)

2 – 3 (*Aa*) and 1 (*A*).

Because diploid progeny were examined in the cross with *aa,* the haploid gametic ratio would be 2*A*:1*a*, and the diploid ratio would also be 2 wild-type : 1 mutant. The table indicates that *y* is on chromosome 1, *cot* is on chromosome 7, and *h* is on chromosome 10.

22. Radiation could have caused point mutations or induced recombination, but nondisjunction is the more likely explanation.

23. P a + c + e × + b + d +

Selection for + + + + +

Because this rare colony gave rise to both parental types among asexual (haploid) spores, the best explanation is that the rare colony initially contained both marked chromosomes due to nondisjunction. That is, it was disomic. Subsequent mitotic nondisjunction yielded the two parental types, possibly because the disomic was unstable.

24. The two chromosomes are

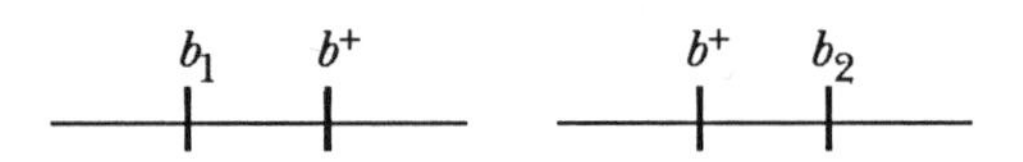

If one of the centromeres becomes functionally duplex before meiosis I, the homologous chromosomes will separate randomly. This will result in one daughter cell having one chromosome with one chromatid, and the second daughter cell will have one chromosome with one chromatid and a second chromosome with two chromatids. Meiosis II will lead to a nullisomic (white) and a monosomic (buff) from the first daughter cell and a disomic (black) and a monosomic (buff) from the second daughter cell.

If both centromeres divide prematurely, each daughter cell will get two chromosomes, each with one chromatid. All the ascospores will be buff.

If nondisjunction occurred at meiosis I, the result will be one-half black (disomic) and one-half white (nullisomic) spores. Nondisjunction at meiosis II for one of the cells will result in two buff spores (normal meiosis II) and one white (nullisomic) and one buff (two copies of the same gene).

Normal meiosis will yield all buff spores because no crossing-over occurs between the two genes.

25. The triploids underwent the equational division at meiosis I and reduction division during meiosis II.

26. a. The white colonies have lost several alleles: *ad—3*$^+$, *leu—1*$^+$, *cyh—2*$^+$. The best interpretation is that the entire chromosome was lost. Because *ad—3* is an earlier block than *ad-2,* the red pigment will not accumulate in the white colonies.

b. Spores from the white colonies will be one-half *ade*$^-$, *leu*$^-$, *cyh*^{r}, *met*$^+$ and one-half nonviable because of the missing chromosome.

27. (A). mutation; (B). mitotic crossover between *w* and *cnx*; (C). nondisjunction

28. Aneuploidy is the result of nondisjunction. Therefore, any system that will detect nondisjunction will work. One of the easiest is that discussed in Problem 15. Cross white-eyed females with red-eyed males and look for the reverse of X-linkage in the progeny. Compare populations unexposed to any environmental pollutants with those exposed to different suspect agents.

Self-Test

1. Prior to meiosis I, a germ cell in a human male experienced the translocation of one entire chromosome 21 to a chromosome 15. What are the chromosome complements of each resulting sperm? If each sperm fertilizes a normal egg, what is the outcome?

2. In *Drosophila,* there is a recessive gene, *ey,* on the fourth chromosome that leads to an eyeless phenotype. A male trisomic for chromosome 4 with the genotype + + *ey* is crossed with an eyeless female. What will be the genotypes and phenotypes of their progeny?

3. How many chromatids are found in prophase of a pentaploid cell when $x = 4$?

4. The diploid number of the mouse is $2n = 44$. How many different trisomics could be formed?

5. A normal couple produces a child with hemophilia, an X-linked recessive disorder. What are the possible genotypes of that child? In which parent did nondisjunction occur?

6. The Abyssinian cat is a tetraploid with 28 chromosomes. Common cat is a hexaploid from the same series. How many chromosomes are in the common cat?

7. A chimpanzee was born at the Yerkes Primate Center in Atlanta that had all the characteristics of Down's syndrome. By analogy with humans, what chromosome abnormality does this chimp have?

Solutions to Self-Test

1. Meiosis I could result in the 15–21 translocation chromosome in one cell and autonomous chromosomes 15 and 21 in the other cell. At meiosis II, each cell would give rise to two identical to it. When they fertilized normal eggs, there would be two completely normal embryos and two embryos that would develop normally but one would be a translocation heterozygote.

Meiosis I could result in disomy 21, with a 15—21 translocation, and nullisomy 21. Both sperm types would result in nonviable embryos.

2. The cross is *ey ey* X + + *ey*. Gametes from the male will be:

1/6 + +

1/6 *ey*

1/3 +

1/3 + *ey*

The progeny will be:

1/6 ++ *ey*	wild-type	1/3 + *ey*	wild-type
1/6 *ey ey*	eyeless	1/3 +ey ey	wild-type

3. (2 chromatids/chromosome)(5 sets of chromosomes/cell)(4 chromosomes/set) = 40 chromatids/cell.

4. 22; all except for the sex chromosomes would very likely be nonviable.

5. The mother must have been heterozygous for hemophilia. An XX female with hemophilia could result from X-inactivation (unlikely) or from nondisjunction in both the male (either division) and the female (meiosis II). The child could be an XO female, with Turner's syndrome. Nondisjunction could have occurred at either division in the male. If the child were an XXY, with Klinefelter's syndrome, nondisjunction at meiosis II in the female occurred. The child could also be a normal XY male, with no nondisjunction involved.

6. If $4x = 28$, $x = 7$ and $6x = 42$.

7. The chimp had trisomy of its smallest autosome.

CHAPTER

10

Recombination in Bacteria and Their Viruses

Important Terms and Concepts

The **prokaryotes** are composed of **blue-green algae** (**cyanobacteria**) and **bacteria**. **Bacteriophages**, or **phages**, are viruses that reproduce in bacteria.

Conjugation is the one-way transfer of DNA from one bacterium to another. The ability to transfer DNA by conjugation is dependent upon the presence of the **F**, or **fertility, factor**.

The F factor is an **episome**, a genetic particle that can exist either free in the cytoplasm or integrated into the host chromosome.

Cells carrying the F factor are F^+ (free in cytoplasm), **Hfr** (integrated into bacterial chromosome), or F' (F factor in cytoplasm, with chromosomal genes inserted into it). Cells lacking the F factor are F^-.

Cells with the F factor produce **pili**, which are proteinaceous structures that attach to the F^- cell. Cells with the F factor also produce a **conjugation tube**, through which genes are transferred.

F^+ cells transfer genes located on the F factor. Hfr cells transfer chromosomal genes and, rarely, the F factor. F' cells transfer both chromosomal and F factor genes; the process is called **sexduction.**

After the transfer of genes, recombination can occur between the transferred genes (**exogenote**) and the F^- **endogenote**. There is a **gradient of transfer** of chromosomal genes.

Transformation is the process by which cells take up naked DNA from their environment. The DNA subsequently recombines into the bacterial recipient.

Phage infection of a bacterial cell can result in cell destruction, **lysis**, or a change in the bacterial cell characteristics, **lysogeny**. Lysis results in a **plaque**, or clear area, in the bacterial lawn.

A **virulent** phage cannot integrate into the bacterial chromosome and therefore always cause lysis. A **temperate** phage can integrate into the bacterial chromosome and therefore cause lysogeny. When integrated, the virus is called a **prophage**.

Phage characteristics include plaque morphology, host range, and burst size.

Transduction is the movement of genes from one bacterial cell to another with the phage as the vector. **Generalized** transduction results from lysis, frequently without a preceding lysogeny; each bacterial gene has an equal chance of being transduced. **Specialized** transduction results from lysogeny; genes close to the site of prophage insertion are transduced.

Be sure that you have thoroughly read the entire chapter before you attempt any of the problems.

Solutions to Problems

1. While the interrupted-mating experiments will yield the gene order, it will only be relative to fairly distant markers. Thus, the precise location cannot be pinpointed with this technique. Generalized transduction will yield information with regard to very close markers, which makes it a poor choice for the initial experiments because of the massive amount of screening that would have to be done. Together, the two techniques allow, first, for a localization of the mutant (interrupted-

mating) and, second, for a precise location of the mutant (generalized transduction) within the general region.

2. This problem is analogous to forming long gene maps with a series of three-point testcrosses. Write the four sequences so that they overlap:

$$
\begin{array}{lllllllll}
\overline{\text{M Z X}} & \text{W} & \text{C} & & & & & & \\
 & \text{W} & \text{C} & \text{N} & \text{A} & \text{L} & & & \\
 & & & & \text{A} & \text{L} & \text{B} & \text{R} & \text{U} \\
 & & & & & & \text{B} & \text{R} & \text{U}\ \underline{\text{M Z}}
\end{array}
$$

The regions with the bars above or below are identical in sequence. The only possible interpretation is that the donated material was circular, with the F factors inserted at different points in different strains.

3. An F⁻ strain will respond differently to an F⁺ (L) or an Hfr (M) strain. Thus strains 2, 3, and 7 are F⁻. Strains 1 and 8 are F⁺, and strains 4, 5, and 6 are Hfr.

4. a.

Agar type	Selected genes
1	c^+
2	a^+
3	b^+

b. The order of genes is revealed in the sequence of colony appearance. Because colonies first appear on agar type 1, which selects for c^+, *c* must be first. Colonies next appear on agar type 3, which selects for b^+, indicating that *b* follows *c*. Allele a^+ appears last. The gene order is *c b a*. The three genes are roughly equally spaced.

c. In this problem you are looking for cotransfer, which results in no growth because the Hfr strain is d^-. Therefore, the farther a gene is from *d*, the more growth that will occur. From the data, *d* is closest to *b*. It is also closer to *a* than it is to *c*. The gene order is *c b d a*.

d. With no A in the agar, the medium is equivalent to agar type 2, and the first colonies should appear at about 17.5 minutes.

5. Induce mating and select for penicillin resistance. Plate onto minimal medium that has been supplemented with all but the selected marker. For example, if ala^+ is being selected, add all amino acids except alanine. Kill the Hfr cells with T1 phage. The recipient cells that did not receive the ala^+ gene will also die.

You will end up with a series of cultures that have one or more genes transferred. Assume that penicillin resistance lies between *ala* and *glu*. Only ala^+, glu^+, and possibly pro^+ and leu^+ cultures could possibly also contain pen^r. Because the order of transferred genes is known, focus on the ala^+ and glu^+ cultures.

Determine the following percentages of colonies:

$$ala^+\ pen^r\ glu^-$$

$$ala^+\ pen^r\ glu^+$$

This will tell you the m.u. between pen^r and glu^+, which in turn will allow you to determine the distance between ala^+ pen^r.

6. a. Determine the gene order by comparing arg^+ bio^+ leu^- with arg^+ bio^- leu^+. If the order were *arg leu bio*, four crossovers would be required to get arg^+ bio^+ leu^-, while only two would be required to get arg^+ bio^- leu^+. If the order is *arg bio leu*, four crossovers would be required to get arg^+ bio^- leu^+, and only two would be required to get arg^+ bio^+ leu^-. The gene order is *arg bio leu*.

b. The *arg—bio* distance is estimated by the arg^+ bio^- leu^- colony type. RF = 100%(48)/376 = 12.76 m.u.

The *bio—leu* distance is estimated by the arg^+ bio^+ leu^- colony type. RF = 100%(8)/376 = 2.12 m.u.

7. To solve this problem, draw the Hfr and recipient chromosomes in both crosses and note the number of crossovers needed to get Z_1^+ Z_2^+ for the two possible gene orders.

Order 1:

Hfr Z_1^- Z_2^+ ade^+ str^s

recipient Z_1^+ Z_2^- ade^- str^r

Order 2:

Hfr Z_2^- Z_1^+ ade^+ str^s

recipient Z_2^+ Z_1^- ade^- str^r

From the number of crossovers required to get Z+ ade- strr, the order must be ade Z2 Z1 str.

8. In crosses A and B, the only types that will grow are pro^+ (*lac—*x^+ *lac—*y^+) *ade+*. Both crosses require a crossover between the *pro* and the *lac* genes, and between the *lac* genes and the *ade* gene. In cross A, if *x* is to the left of *y*, two crossovers are required, and if *y* is to the left of *x*, four crossovers are required. The opposite is true for cross B. Double crossovers are more frequent than quadruple crossovers.

X	Y	Cross A	Cross B	Conclusion
1	2	173	27	1 is to the left of 2.
1	3	156	34	1 is to the left of 3.
1	4	46	218	4 is to the left of 1.
1	5	30	197	5 is to the left of 1.
1	6	168	32	1 is to the left of 6.
1	7	37	215	7 is to the left of 1.
1	8	226	40	1 is to the left of 8.
2	3	24	187	3 is to the left of 2.
2	8	153	17	2 is to the left of 8.
3	6	20	175	6 is to the left of 3.
4	5	205	17	4 is to the left of 5.
5	7	199	34	5 is to the left of 7.

The sequence is *pro*—4—5—7—1—6—3—2—8—*ade.*

9. The most straightforward way would be to put an Hfr at both ends of the same sequence and measure the time of transfer between two specific genes. For example,

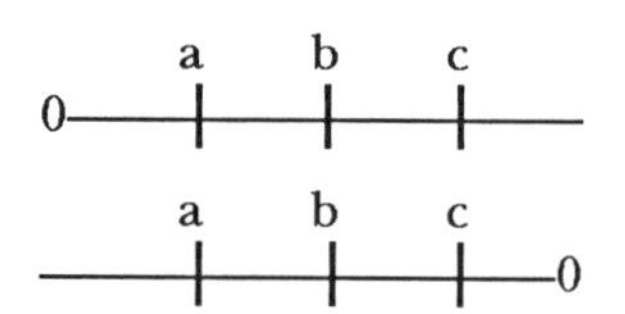

10. a. To survive on the selective medium all cultures must be *ery*^{r}. Keep in mind that 300 of these cells were tested under four separate conditions.

If 263 colonies can grow when only ery is added, they must be *arg*$^+$ *aro*$^+$ *ery*^{r}. The remaining 37 cultures are mutant for one or both genes.

Cultures that can grow on ery + arg are *ery*^{r} *aro*$^+$. They may or may not be *arg*$^+$. Those that cannot grow are *aro*$^-$. Therefore, 300 – 264 = 36 are *ery*^{r} *aro*$^-$. They may or may not be *arg*$^+$.

The 290 cultures that grow on ery + aro must be *arg*$^+$, and the 10 that could not grow must be *arg*$^-$. They may or may not be *aro*$^+$.

Now this information can be assembled in a table. Write what has been determined experimentally, first:

	arg^{+}	*arg*$^{-}$	
aro^{+}	263	?	264
aro^{-}	?	?	36
	290	10	300

Now the unknown values can be filled in.

263 *ery*^{r} *arg*$^{+}$ *aro*$^{+}$

27 *ery*^{r} *arg*$^{+}$ *aro*$^{-}$

1 *ery*^{r} *arg*$^{-}$ *aro*$^{+}$

9 *ery*^{r} *arg*$^{-}$ *aro*$^{-}$.

b. Recombination in the *aro—arg* region is represented by two genotypes: *aro*$^{+}$ *arg*$^{-}$ and *aro*$^{-}$ *arg*$^{+}$. The frequency of recombination is $100\%(1+27)/300 = 9.3$ m.u.

Recombination in the *ery—arg* region is represented by two genotypes: *aro*$^{+}$ *arg*$^{-}$ and *aro*$^{-}$ *arg*$^{-}$. The frequency of recombination is $100\%(1 + 9)/300 = 3.3$ m.u.

Recombination in the *ery— aro* region is represented by three genotypes: *arg*$^{+}$ *aro*$^{-}$, *arg*$^{-}$ *aro*$^{-}$ and *aro*$^{+}$ *arg*$^{-}$. The frequency of recombination is $100\%(27 + 9 + 1)/300 = 12$ m.u.

c. The ratio is 28 :10, or 2.8:1.0.

11. The best explanation is that the integrated *pro*$^{+}$ was sexducted onto an F′ factor that was transferred into recipients early in the mating process. These cells now carry the F factor and are able to transmit F^{+} in the second cross as part of the F' factor, which still carries *pro*$^{+}$.

12. The high rate of integration and the preference for the same site originally occupied by the sex factor suggest that the F′ contains some homology with the original site. The source of homology could be a fragment of the sex factor or it could be a chromosomal fragment.

13. The first task is to get F^{+} cells that are *str*^{r} ala^{+}. To obtain these, cross Hfr *str*^{s} *ala*$^{+}$ with F^{-} *str*^{r} *ala*$^{-}$ and select for *str*^{r} *ala*$^{+}$. These cells will be F^{-}. Assume that they are also *leu*$^{-}$. Mix them with an Hfr line that is *str*^{s} *leu*$^{+}$ *ala*$^{-}$ and select for *str*^{r} *ala*$^{+}$ *leu*$^{+}$. Some of these cells will be F′. Now these cells can be identified by their inability to mate with F^{+} cells or their ability to mate with F^{-} cells.

14. a. and **b.**

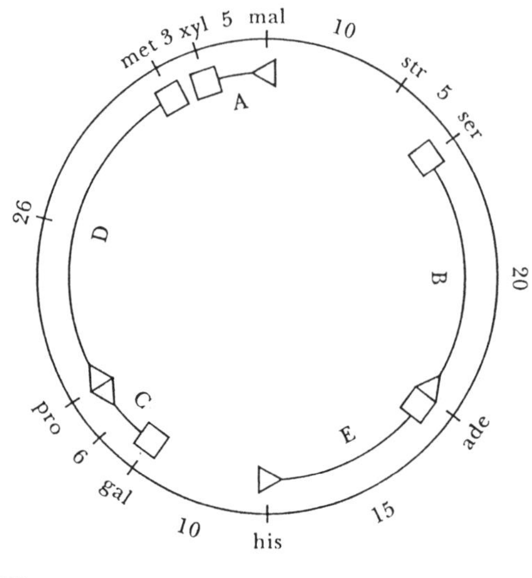

c.

A: select for *xyl* +

B: select for *ser* +

C: select for *gal* +

D: select for *met* +

E: select for *ade* +

15. a. If the two genes are far enough apart to be located on separate DNA fragments, then the frequency of double transformants should be the product of the frequency of the two single transformants, or (4.3%) X (0.40%) = 0.017%. The observed double transformant frequency is 0.17 percent, a factor of 10 greater than expected. Therefore, the two genes are located close together and are cotransformed at a rate of 0.17 percent.

b. Here, when the two genes must be contained on separate pieces of DNA, the rate of cotransformation is much lower, confirming the conclusion in part **a.**

16. a. Notice that each gene was transferred into about one-tenth of the cells (single drugs tested). Also notice that pairwise testing gives low values whenever B is involved but fairly high rates when any drug but B is involved. This suggests that B resistance is not close to the other three genes and it is unlikely that B resistance is brought into the cell on the same piece of DNA as the other three genes.

b. To determine the relative order of genes for resistance to A, C, and D, notice that the frequency of resistance to AC is approximately the frequency of resistance to ACD. Also notice that AD resistance is roughly 50 percent higher. This suggests that D resistance is between the other two genes.

17. The expected number of double recombinants is (0.01)(0.002)(100,000) = 2. Interference = 1 − [(observed DCO)/(expected DCO)] = 1 − 5/2 = −1.5. By definition, the interference is negative.

18. a. m—r: The two parentals are + + + and *m r tu*. The crossovers between *m* and *r* are

m	+	*tu*	162
m	+	+	520
+	*r*	*tu*	474
+	*r*	+	172
			1328

Therefore the distance is 100%(1328)/10,342 = 12.8 m.u.
r—tu: proceeding the same way, the distance is 100%(2152)/10,342 = 20.8 m.u.
m—tu: as above, the distance is 100%(2812)/10,342 = 27.2 m.u.

b. Because *m* and *tu* are farthest apart, the sequence is *m r tu*. At this point, the distance between *m* and *tu* can be corrected for double crossovers (classes + *r* + and *m* + *tu*). The final *m—tu* distance is the sum of the two smaller distances, or 12.8 + 20.8 = 33.6.

c. Recall that I = 1 − c.c. = 1 − (observed DCO/expected DCO). The observed DCO is 162 + 172 = 334. The expected DCO would be (0.128)(0.208)(10,342) = 275. c.c. = 1.2. I = 1 − 1.2 = −.2. A negative value for I indicates that the occurrence of one crossover makes a second crossover more likely to occur than it would have been without that first crossover. That is, more double crossovers occur than are expected.

19. a. I: minimal plus proline and histidine

II: minimal plus purines and histidine

III: minimal plus purines and proline

b. The order can be deduced from cotransfer rates. It is *pur—his—pro.*

c. The closer the two genes, the higher the rate of cotransfer. *His* and *pro* are closest.

d. *Pro*$^+$ transduction requires a crossover on both sides of the *pro* gene. Because *his* is closer to *pro* than *pur*, you get the following:

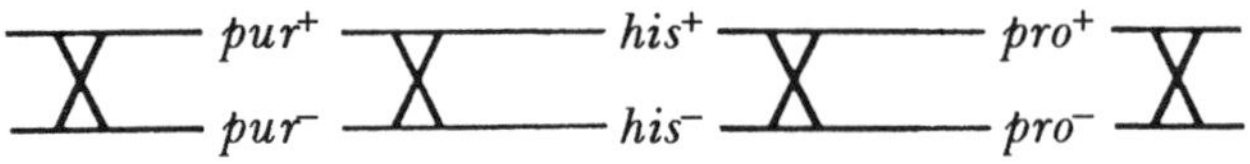

```
   ──── pur+ ─────── his+ ─────── pro+ ───
     X         X           X           X
   ──── pur− ─────── his− ─────── pro− ───
```

A *pur*$^+$ *his*$^-$ *pro*$^+$ genotype requires four crossovers.

20. The instability of *gal*$^+$ transductants comes from the fact that they are partial diploids and have a tendency to lose a *gal* gene. If the *gal*$^+$ gene is lost, a *gal*$^-$ clone will develop. The stable transductants are not partial diploids and, therefore, do not have a tendency to segregate *gal*. They are not partial diploids because there was an exchange of *gal* alleles rather than an insertion of the second *gal* allele, as with partial diploids. This occurred when lambda looped out between the two *gal* alleles and one *gal*$^+$ allele was reconstructed.

21. a. Specialized transduction is at work here. It is characterized by the transduction of one to a few markers.

b. The prophage is located in the *cys—leu* region, which is the only region that gave rise to colonies when tested against the six nutrient markers.

22.

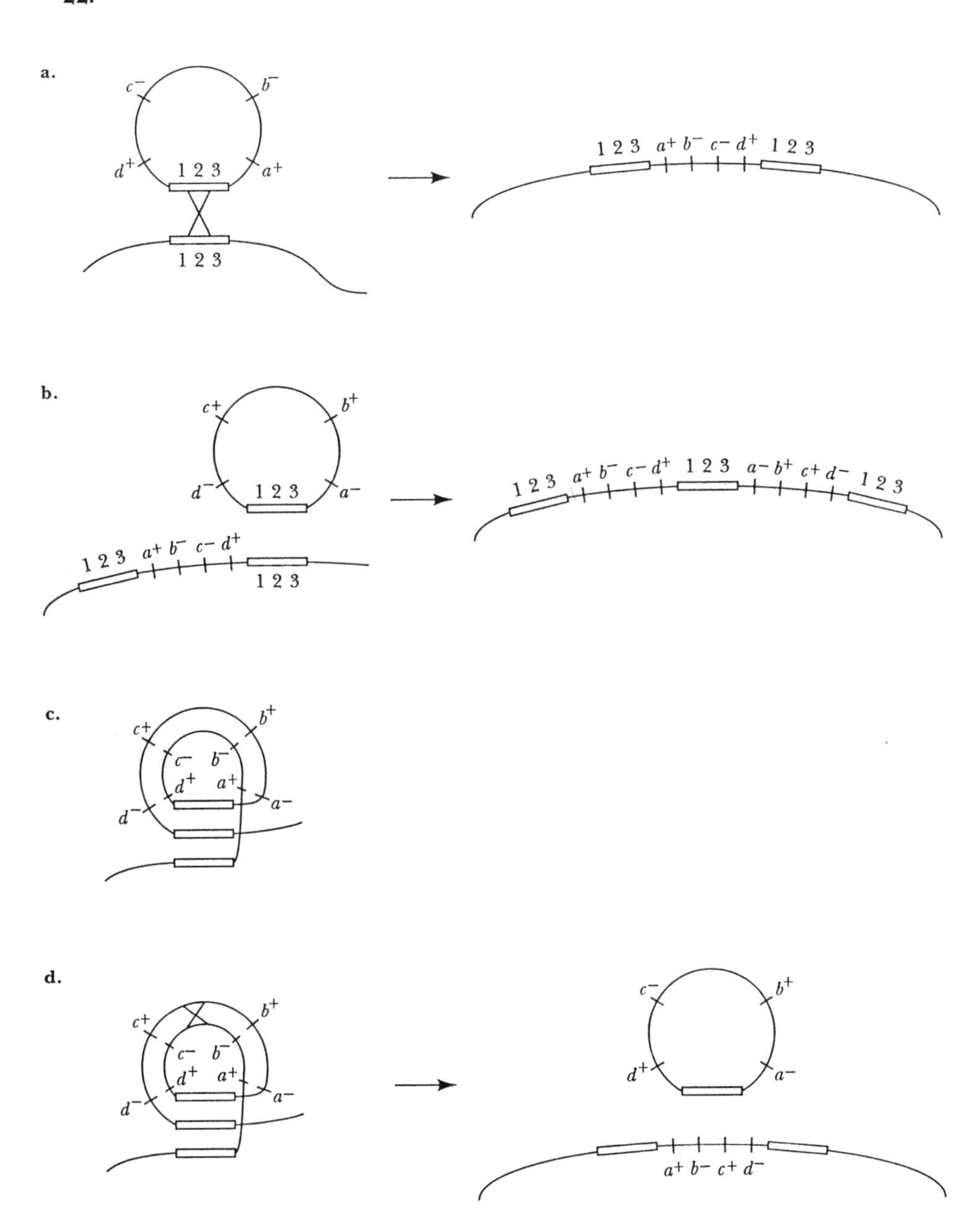

23. a. If *trp1* and *trp2* are alleles, then a cross between strains A and B will never result in trp$^+$, unless recombination occurs within the *trp* gene. If they are not allelic, there will be *trp*$^+$ colonies. Check these colonies for the possibility of recombination.

b. Infect strain C with the Z phage and use the progeny to infect strain B cells (which are immune because they are lysogenic for Z). The strain B cells should be plated onto minimal medium and minimal medium plus cysteine.

If the order is *cys trp2 trp1*, two crossovers will result in cys^+ $trp2^+$ $trp1^+$, and the number of colonies on the two media should be approximately the same.

If the order is *cys trp1 trp2*, four crossovers are required for cys^+ $trp1^+$ $trp2^+$. Therefore, the number of colonies on medium containing cysteine would be greater than the number of colonies on minimal medium.

24. Recognize that if a compound is not added and growth occurs, the *E. coli* has received the genes for it by transduction. Thus, the BCE culture must have received a^+ and d^+. The BCD culture received a^+ and e^+. The ABD culture received c^+ and e^+. The order is thus *d a e c*. Notice that *b* is never cotransduced and is therefore distant from this group of genes.

25. a. $100\%(3 + 10)/50 = 26\%$

b. $100\%(10 + 13)/50 = 46\%$

c. *Pdx* is closer as determined by cotransduction rates.

d. If the order is *pur pdx nad*, four crossovers are required to get pur^+ pdx^+ nad^+. If the order is *nad pur pdx*, two crossovers are required to get wild-type.

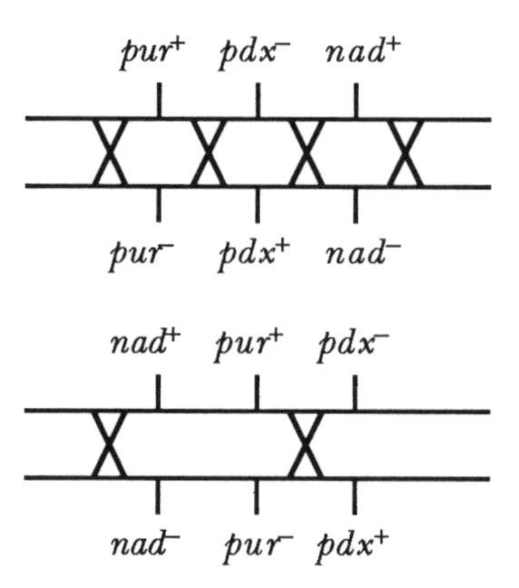

Because completely wild-type occurred less frequently than the other gene combinations, the order is *pur pdx nad*.

26. a. The colonies are all cys^+ and either + or − for the other two genes.

b. (1) cys^+ leu^+ $thr^{+/-}$

(2) cys^+ $leu^{+/-}$ thr^+

(3) *cys* $^+$ *leu*$^+$ *thr* $^+$

c. Because none grew on minimal medium, no colony was *leu*$^+$ *thr* $^+$. Therefore, medium (1) had *cys* $^+$ *leu*$^+$ *thr* $^-$, and medium (2) had *cys*$^+$ *leu*$^-$ *thr*$^+$. The remaining cultures were *cys* $^+$ *leu*$^-$ *thr* $^-$, and this genotype occurred in 100% – 56% – 5% = 39% of the colonies.

d.

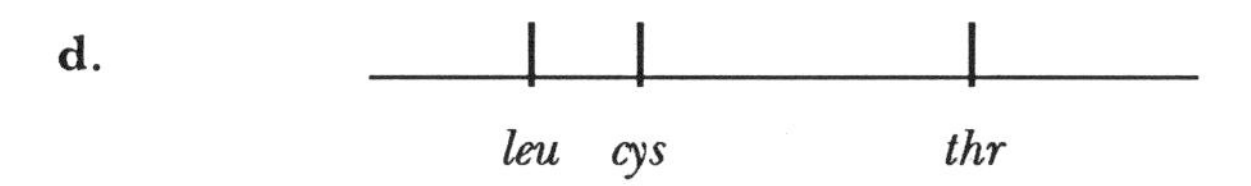

27. To isolate the specialized transducing particles of phage ϕ80 that carried *lac*$^+$, the researchers would have had to induce the phage.

Self-Test

1. Could transformation also occur in eukaryotic cells? What techniques would you use?

2. Could transduction also occur in eukaryotic cells? What techniques would you use?

3. Four Hfr strains donate the markers shown in the order given below. All the Hfr strains are derived from the same F$^+$ strain. What is the order of these markers in the original F$^+$?

Strain 1:	Q	W	D	M	T
Strain 2:	A	X	P	T	M
Strain 3:	B	N	C	A	X
Strain 4:	B	Q	W	D	M

4. A mutant strain is plated onto a complete medium. Replica plates that have minimal medium supplemented by various amino acids are prepared from the original plate.

a. From the results below, determine the genotype of the mutant.

b. Determine the genotype of each of the colonies found.

Replica 1: minimal medium + arg and lys → 1 colony

Replica 2: minimal medium + arg and ser → 0 colonies

Replica 3: minimal medium + lys and ser → 1 colony

5. A cross is made between an Hfr a^+ b^+ c^+ and an *E. coli* strain that is a^- b^- c^-. It is known that *c* enters the recipient last, so cells are selected only for c^+. The exconjugants are tested for a^+ and b^+. The results follow below.

$a^+ b^+ c^+$	300	$a^+ b^- c^+$	3
$a^- b^+ c^+$	0	$a^- b^- c^+$	80

a. What is the gene order?

b. What are the map distances in recombination units?

6. In a generalized transducing experiment, the donor was $a^+ b^+ c^+$ and the recipient was $a^- b^- c^-$. The a^+ was selected after transduction. Sixty one a^+ transductants were then tested for the other genes.

a. What is the cotransduction frequency for *a* and *b*?

b. What is the cotransduction frequency for *a* and *c*?

c. Which locus is closer to *a*?

d. What is the gene order?

Solutions to Self-Test

1. Transformation is defined as the uptake of naked DNA from the environment, followed by a recombination event that leads to change in cellular characteristics. The transformation of eukaryotic cells is routinely utilized in somatic cell genetics. When intact chromosomes are utilized, the process is known as chromosome-mediated-gene-transfer (CMGT). When naked DNA is utilized, the process is known as DNA-mediated-gene-transfer (DMGT).

2. Transduction is the transfer of genetic material from one cell to another, mediated by a virus vector. A form of generalized transduction is used in somatic cell genetics. A virus particle is fractionated into genetic material and the protein from the viral capsid. Naked DNA from the cells selected as donor is mixed with the viral protein. The hybrid viral particle reconstitutes itself by a process known as self-assembly. These hybrid viruses can now be used to transduce cells.

3. Q W D M T P X A C N B Q

4. a. lys^- ser^- arg^-, a triple mutant

b.replica 1: lys^- ser^+ arg^-

replica 3: lys^- ser^- arg^+

5. a. *c a b*

b. c—a: 100%(80)/383 = 20.9 m.u.

a—b: 100%(3)/383 = 0.78 m.u.

6. a. $(a^+ b^+)/a^+ = 16/61 = 26.2\%$ cotransduction

b. $(a^+ c^+)/a^+ = 34/61 = 55.7\%$ cotransduction

c. c is closer to a

d. The order is a c b.

CHAPTER

11

The Structure of DNA

Important Terms and Concepts

The **transforming principle** is DNA.

DNA is made of four **nucleotides.** A nucleotide contains a phosphate group, a deoxyribose sugar, and one of four **bases.** A **nucleoside** contains a deoxyribose sugar and one of four bases. The bases are **adenine, guanine, cytosine,** and **thymine.** Adenine and thymine are **purines**; guanine and cytosine are **pyrimidines.** The number of thymine bases plus cytosine bases equals the number of adenine bases plus guanine bases.

DNA is a right handed **double helix.** The nucleotides are connected by **phosphodiester bonds,** and the chains of the helix are held together by **hydrogen bonds.** The two chains are **antiparallel.**

Replication of the DNA is **semiconservative.** Each chain acts as a **template** during replication. Replication proceeds from the **origin bidirectionally. DNA polymerase** catalyzes the reactions that lead to **polymerization. DNA ligase** ligates two DNA molecules together. **DNA**

topoisomerase converts DNA from one topological form to another. **DNA gyrase** can induce twisting and coiling of DNA leading to **supercoiling.** The bacterial **primase** synthesizes an RNA primer for DNA polymerase.

Each chromatid is composed of a single DNA helix.

Be sure that you have thoroughly read the entire chapter before you attempt any of the problems.

Solutions to Problems

1. Because A = T, G = C, and A + T + G + C = 1, 1 – 2T = G + C. Therefore, 1 – 2T = 2C. C = 1/2(1 – 2T) = 1/2 (1 – 0.30) = 35 percent.

2.

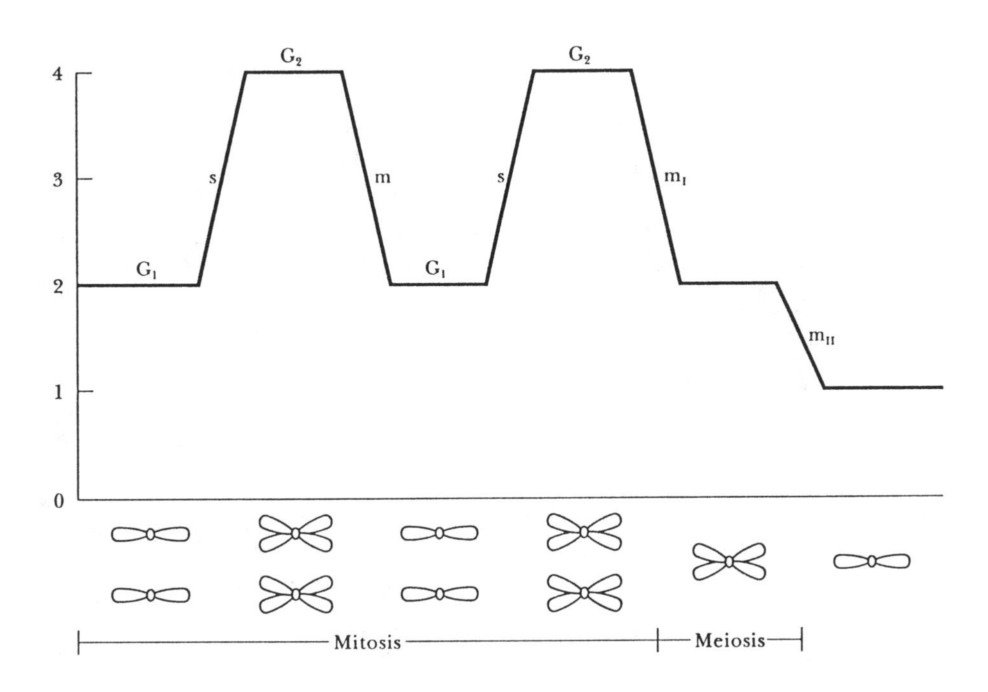

3.

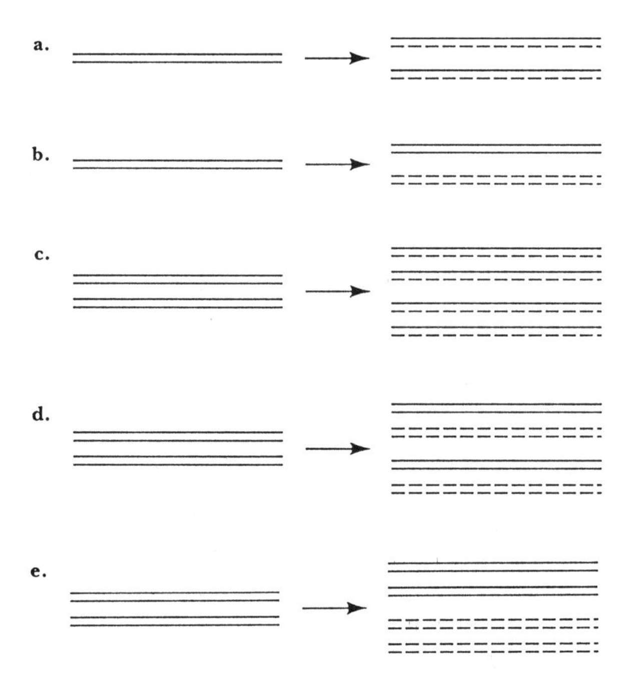

f. Models b and e are ruled out by the experiment. The results were compatible with semiconservative replication, but the exact structure could not be predicted from the results.

4. The results suggest that the DNA is replicated in short segments that are subsequently joined by enzymatic action (ligase). Because DNA replication is bidirectional, because there are multiple points along the DNA where replication is initiated, and because DNA polymerase work only in a 5'→3' direction, one strand of the DNA is always in the wrong orientation for the enzyme. This requires synthesis in fragments.

5. Replication requires that the enzymes and initiation factors of replication have access to the DNA. One possible answer is that the heterochromatic regions, being more condensed than euchromatin, require a longer period to decondense to the point where replication can be initiated. A second possibility is that the heterochromatic regions, which are generally located at centromeres and telomeres in many organisms, act as "anchors" to the nuclear membrane for the chromosome. Embedded in part in the nuclear membrane, these regions require extra time to disentangle so that replication may proceed. A third possibility involves the mechanism of heterochromatization. Genes that

are inactive in a given cell type are generally heterochromatic. This mechanism may also specifically delay replication. Many other possibilities can be proposed.

6. a. A very plausible model is of a triple helix, which would look like a braid, with each strand interacting by hydrogen bonding to the other two.

b. Replication would have to be terti-conservative. The three strands would separate, and each strand would dictate the synthesis of the other two strands.

c. The reductional division would have to result in three daughter cells, and the equational would have to result in two daughter cells, in either order. Thus, meiosis would yield six gametes.

7. Chargaff's rules are that A = T and G = C. Because this is not observed, the most likely interpretation is that the DNA is single stranded. The phage would first have to synthesize a complementary strand before it could begin to make multiple copies of itself.

8. Remember that there are two hydrogen bonds between A and T, while there are three hydrogen bonds between G and C. Denaturation involves the breaking of these bonds, which requires energy. The more bonds that need to be broken, the more energy that must be supplied. Thus the temperature at which a given DNA molecule denatures is a function of its base composition. The higher the temperature of denaturation, the higher the percentage of GC pairs.

9. Deletions and duplications at the DNA level would look exactly like the same events during homologous pairing of chromosomes at the duplex DNA level:

Inversions would have two possibilities, depending on the relative length of the inverted and noninverted regions:

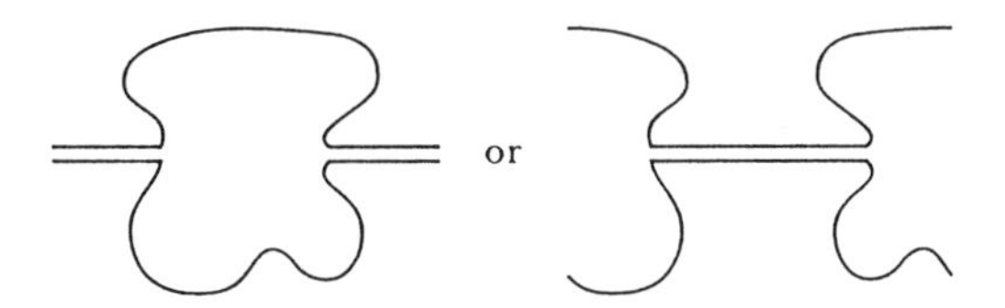

10. a. The first shoulder appears before strand interaction takes place, suggesting that the complementary regions are in the same molecule. This is called a palindrome:

ATGCATGGCCA—TGGCCATGCAT

TACGTACCGGT—ACCGGTACGTA

When the strands separate, each strand can base-pair with itself:

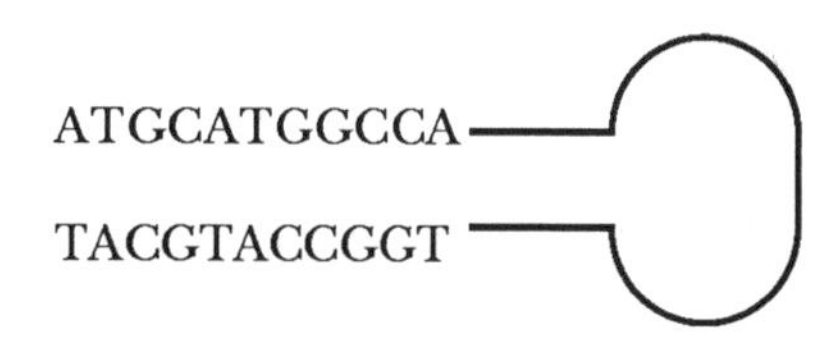

b. The second shoulder represents sequences that are present in many copies in the genome. Because they are at a higher concentration than unique sequences, they have a higher probability of encountering each other during a given time period.

11. First realize that the repeated and unique sequences could be completely interspersed, partially interspersed, or completely segregated from each other. The distribution of these two types of sequences will affect the C_0T curves observed.

Extract the DNA from the cells. Do separate renaturation curves for unsheared DNA and DNA sheared to just a few kilobases.

If there is generalized dispersion, then there will be a shift to the left for unique sequences in the unsheared sample because the interspersed repeated sequences will facilitate pairing and because there are fewer unique sequences in any given segment.

You might also examine the rapidly annealing DNA from the unsheared sample by electron microscopy. Interspersion would result in partial duplexes, with unmatched single stranded regions (unique sequence DNA) interspersed.

12. This observation suggests that the mouse cancer cells have copies of the virus genome integrated into them. Thus, it may be that the viral genes somehow alter cell function, triggering malignancy. Alternatively, the viral genome may carry one or more genes that directly result in malignancy. In either case, viral infection may also be the mechanism by which human malignancy is triggered.

13. The data suggest that each chromosome is composed of one long, continuous molecule of DNA.

14. Let the broken line indicate DNA that has incorporated bromodeoxyuridine and the unbroken line indicate normal DNA.

After one round: After two rounds: After three rounds:

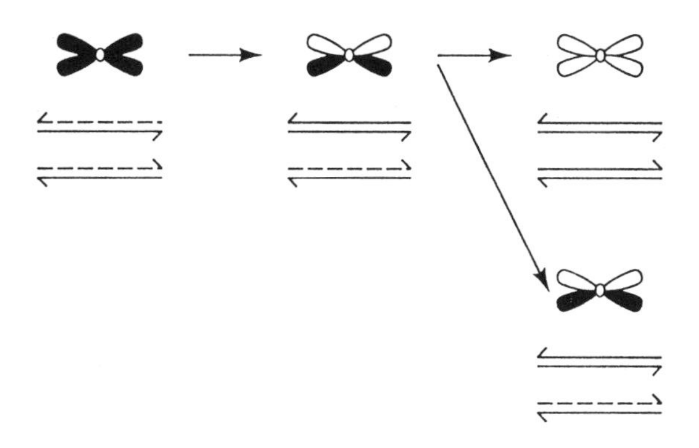

Self-Test

1. Which of the following two sequences would you be more likely to find in an organism that grows in a hot spring? Why?

a. 3′ATATCTGATTTAT5′

b. 3′GGGCGTGGGCGGA5′

2. By most genetic measures, chimpanzees and gorillas are closer to each other than humans are to either. Recently, DNA hybridization was used as a measure of genetics relatedness, and the results indicated that humans and chimpanzees are the closest pair. The results were confirmed independently. This conclusion has been hotly contested by a number of researchers. What are their possible arguments?

3. If a DNA sequence is 3′ATTGC5′, what is its complementary sequence?

4. Assume that a single break occurred in a DNA chain. Which enzyme is capable of connecting the two pieces?

5. Buoyant density is a measure of molecular density during centrifugation. The more dense a molecule, the faster it will pellet in a given solution. The buoyant density (ρ) of DNA molecules in 6M CsCl increases with the molar content of G + C nucleotides according to the following formula:

$$\rho = 1.660 + 0.00098\ (G + C)$$

Find the molar percentage of G + C in DNA from *Streptococcus pneumoniae*, in which $\rho = 1.700$. What is the A + T molar content?

6. Which enzyme initiates transcription in bacteria?

7. Which enzyme fills gaps during DNA replication?

8. Which enzyme is the primary DNA replication enzyme?

9. What class of enzymes is responsible for changing DNA topology?

10. Which enzyme unwinds DNA during replication?

Solutions to Self-Test

1. b. The high GC content, with three hydrogen bonds, would be more resistant to melting than a high AT content, with only two hydrogen bonds.

2. The objections have ranged from the highly technical to the highly slanderous. One scientific objection is that DNA hybridization is much more complex than has been recognized. Another scientific objection revolves around the statistical analysis that was conducted. A third is that because the results contradict the results obtained by several other approaches, DNA hybridization simply is not a good measure of genetic relatedness.

3. 3´GCAAT 5´

4. DNA ligase

5. The equation is $1.700 = 1.660 + 0.00098(G + C)$. $G + C = 40.8$ percent.

Because $(A + T) + (G + C) = 1.0$, the A + T molar percentage is 59.2.

6. primase

7. DNA polymerase I

8. DNA polymerase III

9. topoisomerases

10. helicase

CHAPTER

12

The Nature of the Gene

Important Terms and Concepts

Genes control biochemical reactions by controlling the production of enzymes. **Enzymes** are proteins, which consist of one or more chains of **amino acids** linked together by a **peptide bond**. A **polypeptide** is a chain of amino acids. One or more polypeptides exist in each enzyme. A protein-encoding gene specifies the sequence of amino acids in a polypeptide.

A change in base sequence can lead to a change in amino acid sequence, which can result in an alteration of function of the enzyme.

Recombination can occur within a gene. A **recon** is the unit of recombination; it is equal to one nucleotide base pair.

Mutation can occur within a gene. A **muton** is the unit of mutation; it is equal to one nucleotide base pair.

A **cistron** is the equivalent of a gene. It is a genetic region within which normally no complementation occurs between mutations.

Be sure that you have thoroughly read the entire chapter before you attempt any of the problems.

Solutions to Problems

1. The defective enzyme that results in albinism may not be able to detoxify a chemical component of Saint-John's wort that the wild-type enzyme can detoxify.

2. A secondary cure would result if all lactose is removed from the diet. The disorder would be expected not to be dominant, because one good copy of the gene should allow for at least some, if not all, breakdown of lactose. In fact, the disorder is recessive.

3. Amniocentesis can be used to detect chromosomal abnormalities and any single gene abnormalities for which there is a test. Thus, Down's syndrome, Turner's syndrome, and other chromosomal abnormalities can be detected. Furthermore, biochemical disorders such as galactosemia, Tay-Sachs disease, sickle cell anemia, and phenylketonuria can be detected. Amniocentesis would thus be useful whenever there is a family history of chromosomal or biochemical disorders, whenever there is a history of parental exposure to mutagens (X rays or chemicals), and when the maternal age is over 35, because the rate of nondisjunction is elevated in this population.

4. **a.** The main use is in detecting carrier parents and in diagnosing a disorder in the fetus.

b. Because the values for normal individuals and carriers overlap for galactosemia, there is ambiguity if a person has 25 to 30 units as a result. That person could be either a carrier or normal.

c. These genes are phenotypically dominant but are incompletely dominant at the molecular level. A minimal level of enzyme activity apparently is enough to ensure normal function and phenotype.

5. One less likely possibility is a germ-line mutation. More likely is that each parent was blocked at a different point in a metabolic pathway. If one were *AA bb* and the other were *aa BB*, then the child would be *Aa Bb* and would have sufficient levels of both resulting enzymes to produce pigment.

6. Assuming homozygosity, the children would be normal (see Problem 5).

7. a. Complementation refers to genes within a cell, which is not what is happening here. Most likely what is known as cross-feeding is occurring, whereby a product made by one strain diffuses to another

strain and allows growth of the second strain. This is equivalent to supplementing medium. Because cross-feeding seems to be taking place, the suggestion is that the strains are blocked at different points in the metabolic pathway.

b. For cross-feeding to occur, the growing strain must have a block that occurs earlier in the metabolic pathway than does the block in the strain from which it is obtaining the product for growth.

c. $E - D - B$

d. Without some tryptophan, no growth at all would occur, and the cells would not have lived long enough to produce a product that could diffuse.

8.

Experiment	Result	Interpretation
v into *v* hosts	scarlet	defects in same gene
cn into *cn* hosts	scarlet	defects in same gene
v into wild-type hosts	wild-type	wild-type provides *v* product
cn into wild-type hosts	wild-type	wild-type provides *cn* product
cn into *v* hosts	scarlet	*v* cannot provide *cn* product; *cn* later than *v* in metabolic pathway
v into *cn* hosts	wild-type	*cn* provides *v* product; *v* defect earlier than *cn*

A simple test would be to grind up *cn* animals, inject *v* larvae with the material, and look for wild-type development.

9.

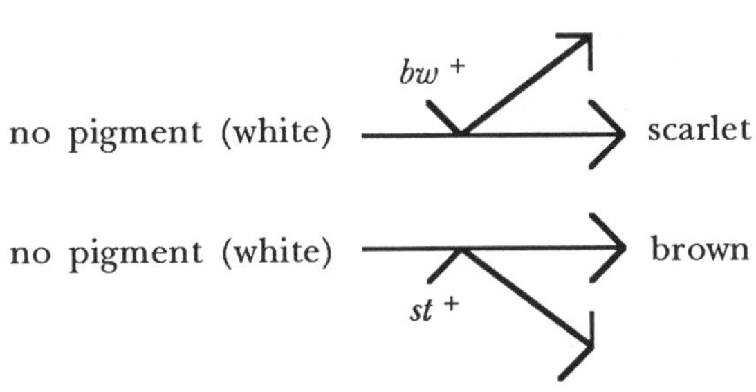

Scarlet plus brown results in red.

10. a. The later a compound is in a pathway, the more mutants for which it can support growth. Therefore,

$$E \rightarrow A \rightarrow C \rightarrow B \rightarrow D \rightarrow G$$

b. A block is indicated by the fact that no growth takes place if a compound is provided that is already being made. Growth occurs if a compound that cannot be made by the organism is supplied.

Mutant 1: grows on D and G; the block is at the conversion of B to D.

Mutant 2: grows on B, D and G; the block is at the conversion of C to B.

Mutant 3: grows on G; the block is at at the conversion of D to G.

Mutant 4: grows on B, C, D and G; the block is at at the conversion of A to C.

Mutant 5: grows on all but E; the block is at at the conversion of E to A.

c. 1,3 + 2,4: 1,3 would accumulate D and require G; 2,4 would require B, D or G but could then make G for 1,3 growth. Therefore, growth would occur.

1,2 + 2,4: 1,2 would require D or G and accumulate B; 2,4 would require B, D, or G and accumulate C. Neither can help the other to grow

1,2 + 2,4 + 1,4: 1,2 would require D or G and would accumulate B; 2,4 would require B, D, or G and accumulate C; 1,4 would require D or G and would accumulate B. Therefore growth.

11. a. When $m = 0.5$, $e^{-m} = 0.60$. Therefore, 60 percent of the meioses will not have a crossover.

b. Because RF = $1/2(1 - e^{-m})$, RF = $1/2(1 - 0.6) = 0.20$. That is, there are 20 m.u. between the two genes.

c. The recombinants will be 1/2(*val—1 val—2*) and 1/2(+ +). Therefore, 1/2RF = + + = 10 percent.

d. Remember that accumulation of substance x means the gene responsible for converting substance x to the next metabolic substance is defective. Also, if substance y permits growth, it is beyond a block in a metabolic pathway.

$$\rightarrow B \xrightarrow{1} A \xrightarrow{2} \text{valine}$$

12. a. A defective enzyme B (from m_2m_2) would yield red petals.

b. Purple, because it has a wild-type allele for each gene.

c. 9 $M_1 - M_2 -$ purple

3 $m_1m_1\ M_2 -$ blue

3 $M_1 - m_2m_2$ red

1 m_1m_1 m_2m_2 white

d. Because they do not produce a functional enzyme.

13. a. white

b. blue

c. purple

d.

P	*bb DD* × *BB dd*
F_1	*Bb Dd* × *Bb Dd*
F_2	9 *B* — *D* — purple
	3 *bb D* — white
	3 *B* — *dd* blue
	1 *bb dd* white or 9:3:4

14. The cis and trans burst size should be the same if the mutants are in different cistrons, and if they are in the same cistron, the trans burst size should be zero. Therefore, assuming *rV* is in *A*, *rW* also is in *A*, and *rU*, *rX*, *rY*, and *rZ* are in *B*.

15. The cross is *pan2x* + X + *pan2y*.

a. If one centromere precociously divides, that will put three chromatids in one daughter cell and one in the other.

daughter cell 1:

pan 2x +
------------O

daughter cell 2:

\+ *pan 2y*
------------O
------------O

pan 2x +
------------O

After meiosis II and mitosis, the first daughter cell would give rise to two pale (pan2x +) and two white aborted (nullisomic) ascospores, while the second daughter cell would give rise to two black (pan2x +/+ pan2y) and two pale (+ pan2y) ascospores. The same result (4 pale : 2 colorless : 2 black) would occur if only the other centromere divided early.

b. If both centromeres divided precociously, each daughter cell would be pan2x +/+ pan2y. This would lead to 4 colorless (nullisomic) and 4 black (disomic) ascospores.

16. a. The mutant does not complement any other mutant. The best interpretation is that it is a deletion.

b. Complementation groups do not complement within a group (–) but do complement between groups (+). Notice that mutants 1, 5, 8, and 9 complement all others but do not complement within the group. The same holds for mutants 2, 3, 4, and 12 as a group and mutants 6, 7, 10, 11, and 13 as a group. These are the three complementation groups.

c. Mutants 1 and 2 are in different cistrons, so the cross can be written 1 + X + 2. Assuming independent assortment, the progeny would be

1/4 1 + *eye*$^-$

1/4 1 2 *eye*$^-$

1/4 + 2 *eye*$^-$

1/4 + + *eye*$^+$

or 3 *eye*$^-$: 1 *eye*$^+$

Mutants 2 × 6 also complement each other. If independent assortment existed, a 3:1 ratio would be observed. Because the ratio is 113:5, there is no independent assortment and the cistrons are linked. Only one of the two recombinant classes can be distinguished: 5 eye$^+$. Because the recombinants should be of equal frequency, the total number of recombinants is 10 out of 118, which leads to 100% (10/118) = 8.47 m.u. distance between the cistrons.

Mutant 14 includes the same cistron (no complementation) as mutant 1.

d. There are three complementation groups; therefore there are three loci, plus mutant 14. Because two of the groups are independently assorting, either mutant 14 is a very large deletion spanning the three loci (and they are therefore on the same chromosome) or it is a separate fourth locus that in some fashion controls the expression of the other three loci or it is a double mutant with a point mutation within one complementation group (1, 5, 8, 9) and a deletion spanning the two linked complementation groups.

e. Two groups are linked, with 8.5 m.u. between them: (2,3,4,12 and 6,7,10,11,13), and the third group is either on a separate chromosome or more than 50 m.u. from the two other groups.

17. The best interpretation is that r_x is a deletion spanning r_c and r_d. Alternatively, it could be a double point mutation in r_c and r_d. If it is a double point mutation, then a cross of r_x with a mutant between r_c and r_d should yield wild-type at a low frequency (double crossover). If it is a deletion, no wild-type will be observed.

$$\text{P} \quad + r_{c-d} + \text{X}\; r_c + r_d\; (r_x)$$

Double recombinants: + + + and $r_c\; r_{c-d}\; r_d$

18. a. A point mutation within a deletion yields no growth (–), while a point mutation external to a deletion yields growth (+). Only mutant *c* fails to complement deletion 1, so it is within the deletion and the other mutants are not. Mutant *d* is in the overlapping region of deletions 2 and 3, while mutant *e* is in the small region of deletion 2 that is not overlapped by other deletions. The partial order is therefore *c e d*. Mutant *a* is within the nonoverlapped deletion 3, and mutant *b* is within deletions 3 and 4. The final map is

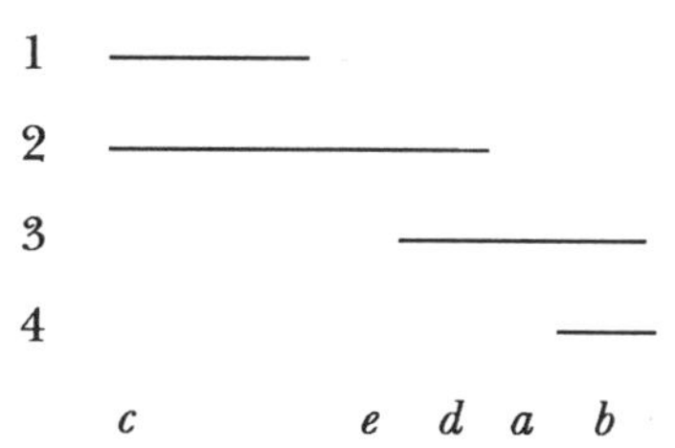

b. The suggestion is that deletion 4 spans both cistrons. This does not affect the conclusions from part **a.**

19. a. Here, + indicates nonoverlapping and – indicates overlapping. Therefore, 1 overlaps 3 and 5, 2 overlaps 5 only, 3 overlaps all but 2, 4 overlaps 3 only, and 5 overlaps all but 4. Putting all these pieces together yields the deletion map

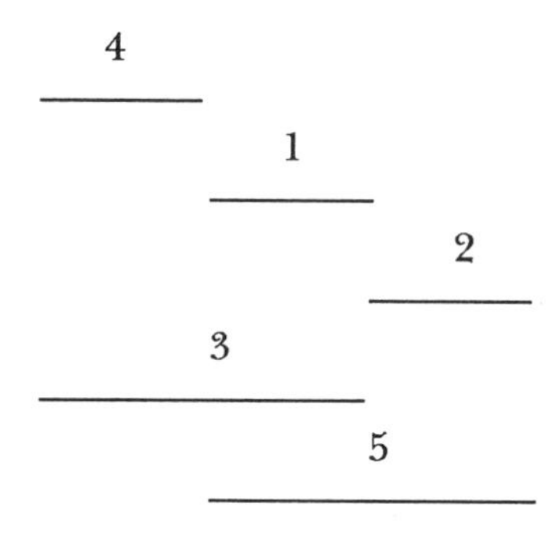

b.

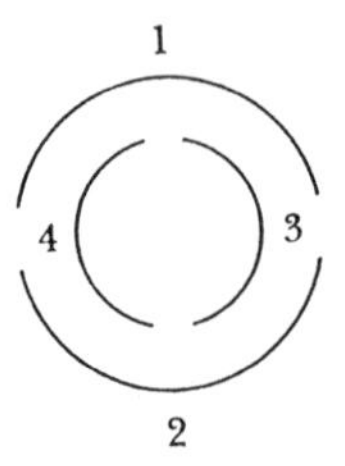

20. a. The values equal one-half of the recombinants. Because the problem asks for relative frequency, however, it is not necessary to convert to real m.u. Therefore, the distances are

1-2: 14	2-3: 12
1-3: 2	2-4: 6
1-4: 20	3-4: 18

The only map possible is

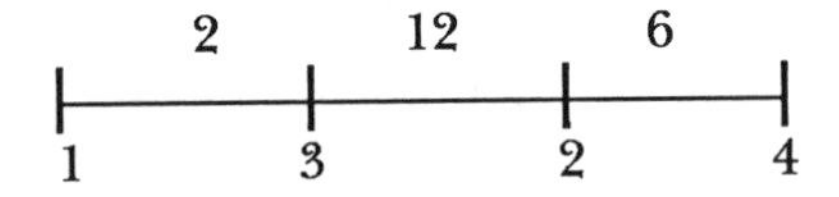

b. no

21. This is like any other mapping problem when the reciprocal cannot be detected: the detected class is multiplied by 2. Rewrite the crosses and results, using parentheses to indicate unknown order, so that it is clear what the results mean.

Cross 1: $his\text{—}2^+\ (a\ b^+)\ nic\text{—}2^+ \times his\text{—}2\ (a^+\ b)\ nic\text{—}2$

Cross 2: $his\text{—}2^+\ (a\ c^+)\ nic\text{—}2 \times his\text{—}2\ (a^+\ c)\ nic\text{—}2^+$

Cross 3: $his\text{—}2\ (b\ c^+)\ nic\text{—}2 \times his\text{—}2^+\ (b^+\ \mathrm{c})\ nic\text{—}2^+$

Results:

	Cross 1	Cross 2	Cross 3
his—2 + nic—2	DCO between *a* and *b*, and *b* and *nic*	1 CO between *a* and *c*	DCO between *b* and *c*, and *b* and *nic*
his—2+ + nic—2+	DCO between *a* and *b*, and his and *a*	as above	DCO between *b* and *c*, and his and *c*

his—2 + nic—2+	1 Co between *a* and *b*	DCO between *a* and *c*, and *nic* and *c*	1 CO between *b* and *c*
his—2+ + nic—2	3 CO between *a* and *b*, *his* and *a*, and *b* and *nic*	DCO between *a* and *c*, *his* and *a*,	3 CO between *b* and *c*, *his* and *c*, and *b* and *nic*

Putting all this information together, the only gene sequence possible is *his—2 a c b nic—2*. To calculate the genetic distances, it is necessary to multiply the frequency of recombinants by 2 because reciprocals are not seen:

$$a\text{-}b\text{: } 100\%(2)(15)/41{,}236 = 0.072 \text{ m.u.}$$

$$a\text{-}c\text{: } 100\%(2)(6)/38{,}421 = 0.031 \text{ m.u.}$$

$$b\text{-}c\text{: } 100\%(2)(5)/43{,}600 = 0.023 \text{ m.u.}$$

The final map is

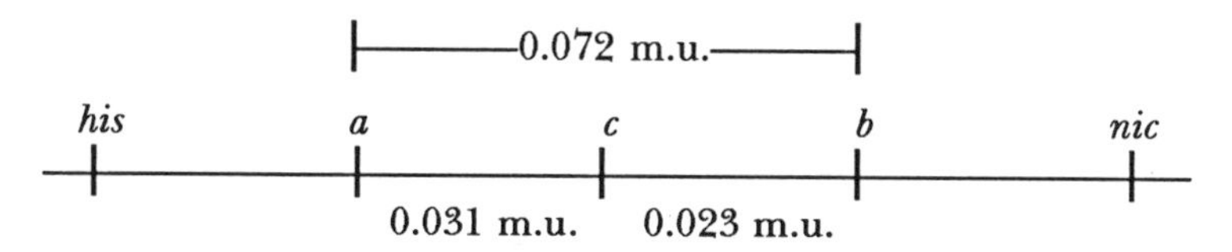

22. Class I mutants are probably deletions, while the class II mutants are probably point mutations. A homozygous deletion would be expected to be lethal, while point mutations would be expected to be recessive. The heterozygote shows both phenotypes because the point mutation can be expressed when paired with a deletion.

To test this interpretation, cross a class I mutant for eye effects with a class II mutant for wings. The heterozygote should show only dominant eye effects, because the class I mutant should be normal for the class II point mutation. Next, cross the heterozygote to another class II mutant, this time for bristles. The progeny should be 1/2 wild-type : 1/2 class I eye mutant.

23. The allele s^n will show dominance over s^f because there will be only 40 units of square factor in the heterozygote. Here, the functional allele is recessive. The allele s^f may become dominant over time in two ways: (1) it could mutate slightly, so that it produces more than 50 units, or (2) other modifying genes may mutate to increase the production of s^f.

24. Pleiotropic effects result from interaction between gene products. As an example, sickle cell anemia results from a point mutation in the globin portion of hemoglobin. The phenotype can include joint damage, brain damage, kidney damage, etc. These effects can be labeled pleiotropic, but because the molecular, cellular, and organismal bases for the damage are understood, the term *syndrome* is usually used. As another example, a gene product (an enzyme) may act in the branch point of a metabolic pathway. The lack of this gene product would then result in two or more deficiencies, and each endpoint effect would be labeled a pleiotropic effect. A third example would be regulatory proteins (Chapters 16, 21, and 22), which can alter many functions simultaneously.

25. a. One mutant cannot utilize the CAP produced by the wild-type allele of the other gene. Thus, there must be two pools of CAP in the cell, and neither pool is available for use in the alternative metabolic pathway. This could occur through membrane sequestering of each reaction.

b. Because the *pyr—3* enzyme controls two consecutive metabolic steps, the suggestion is that CAP is not released by the enzyme. This accounts for the inability of *pyr—3*$^+$ to compensate for *arg—3* mutants, but not vise versa. The *pyr—3* enzyme must have at least two active sites.

c. Partial suppression most probably results from a buildup in CAP to the point where some of it is released by the nonfunctional enzymes. When released, it is available for utilization through complementation.

d. Again, the facts suggest that the biochemical pathways are kept separate.

e. In rats, there are no separate pools of CAP.

26. Benzer used the Poisson distribution to make his calculations. While he knew the number of 1 and 2 occurrence sites, he did not know the total number of occurrence sites. Therefore, he had to assume that the Poisson distribution was applicable.

The equation is $f(i) = e^{-m}m^i/\, i!$.

$$\text{The terms } f(0) = e^{-m}m^0/0! = e^{-m},$$

$$f(1) = e^{-m}m^1/1! = e^{-m}m^1 = 117 \text{ (from Figure 11-35)}$$

$$f(2) = e^{-m}m^2/2! = e^{-m}m^2/2 = 53 \text{ (from Figure 11-35)}$$

These equations can be used to determine the value of m:

$$f(2)/f(1) = (e^{-m}m^2/2)/(e^{-m}m^1) = 53/117$$

$$m/2 = 53/117$$

$$m = 0.9059$$

The number of 0 occurrences is

$$f(0)/f(1) = e^{-m}/e^{-m}m$$

$$f(0) = f(1)/m = 117/0.9059 = 129.15$$

27. *Cross 1 × 2:* All purple F_1 indicates that two genes are involved. Call the defect in 1 *aa* and the defect in 2 *bb*. The cross is

P $\quad aa\ BB \times AA\ bb$

F_1 $\quad Aa\ Bb$

If the two genes assort independently, a 9:7 ratio of purple : white would be seen. A 1:1 ratio indicates tight linkage. The cross above now needs to be rewritten

P $\quad a\ B/\ a\ B \times A\ b/A\ b$

F_1 $\quad a\ B/A\ b$

F_2 $\quad$ 1 $a\ B/a\ B$ (white) : 2 $a\ B/A\ b$ (purple) : 1 $A\ b/A\ b$ (white)

Cross 1 × 3: Again, an F_1 of all purple indicates two genes. The 9:7 F_2 indicates independent assortment. Therefore, let the cross 3 defect be symbolized by *d*:

P $\quad a\ B/a\ B\ \ DD \times A\ B/A\ B\ \ dd$

F_1 $\quad a\ B/A\ B\ \ Dd \times a\ B/A\ B\ \ Dd$

F_2 $\quad$ 9 *A— BB D—* (purple) : 3 *aa BB D—* (white) : 3 *A— BB dd* (white) : 1 *aa BB dd* (white)

Cross 1 × 4: All white F_1 and F_2 indicates that the two mutations are in the same gene. The cross is

P $\quad a\ B/a\ B\ \ DD \times a\ B/a\ B\ \ DD$

F_1 $\quad$ same as parents

F_2 $\quad$ same as parents

Cross 2 × 3:

P $\quad A\ b/A\ b\ \ DD \times A\ B/A\ B\ \ dd$

F_1 $\quad A\ b/A\ B\ \ Dd$ (purple)

F_2 9 *A— B— D—* (purple) : 3 *A b/A b D—* (white) : 3 *A— B— dd* (white)
: 1 *A b/A b dd* (white)

Cross 2 × 4: same as cross 1 × 2

28. Mutants *a* and *e* have point mutations within the same cistron. The other point mutations are all in different cistrons. There are at least four cistrons involved with leucine synthesis.

With the exception of two crosses (*a* × *e*, *b* × *d*), the frequency of prototrophic progeny is approximately 25 percent. This indicates independent assortment of *a* + *e* with *b*, *d*, and *c*, and *c* with *b* and *d*. Cistrons *b* and *d* are linked: RF = 100%(4)/500 = 0.8 m.u.

29. a. There are three cistrons:

Cistron 1: mutants 1, 3, and 4

Cistron 2: mutants 2 and 5

Cistron 3: mutant 6

b. Diagram the heterozygotes that yield $star^+$ gametes, using parentheses to indicate unknown order:

1—6: $A\ (1^+\ 6)\ B/a\ (1\ 6^+)\ b \rightarrow a\ (1^+\ 6^+)\ B$

2—4: $A\ (2^+\ 4)\ B/a\ (2\ 4^+)\ \mathrm{b} \rightarrow a\ (2^+\ 4^+)\ B$

To determine cistron order within the 1—6 cross, note that the order *A* 1 6 *B* would require three crossovers, while the order *A* 6 1 *B* would require one crossover. The order is more likely *A*/*a* 6 (1, 3, 4) *B*/*b*.

To determine cistron order from the 2—4 cross, note that the order *A* 2 4 *B* would require three crossovers, while the order *A* 4 2 *B* would require one crossover. The order is more likely *A*/*a* (4, 1,3) (2, 5) *B*/*b*. The final order is *A*/*a* 6 (1, 3, 4) (2, 5) *B*/*b*.

Self-Test

1. The following point mutations were crossed with the following deletions. In the following table, + indicates complementation, – indicates no complementation. Order the genes along the deletion map.

Point mutations	Deletions A	B	C	D	E
1	–	+	+	+	+
2	+	–	+	–	+
3	+	–	–	+	+
4	+	+	–	+	+
5	–	+	–	+	–

2. Construct a hypothetical biochemical pathway to explain a 10:6 ratio.

3. Several mutants are isolated that can grow when they are supplied with substance D. The mutants were tested for growth when provided with precursors of D. In the following table, + indicates growth and – indicates no growth. What is the order of compounds in the metabolic pathway?

Mutants	A	B	C	D
1	–	–	+	+
2	+	–	+	+
3	–	–	–	+

4. Two deaf parents gave rise to a normal-hearing child. What is the minimum number of cistrons involved?

5. Two mutants known to be in the same cistron were able to complement each other. How can you explain this?

Solutions to Self-Test

1.

```
     ———————— A ————
                ——————— C ———————
                               ———————— B ————————
        ————— E —————            ————— D —————
   1           5     4     3          2
```

2. In a dihybrid cross, let the recessive alleles for both genes produce a functional enzyme.

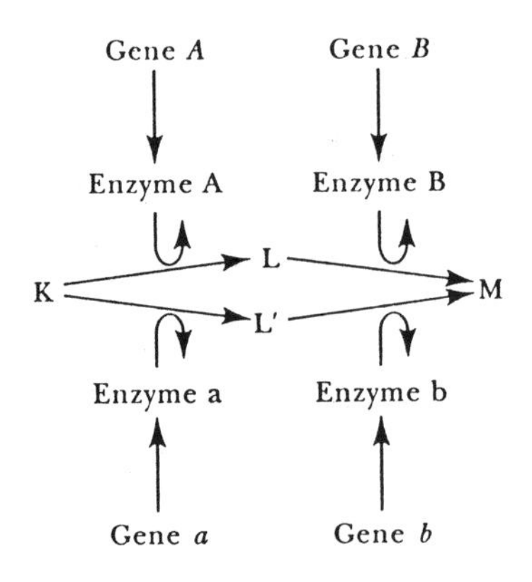

genotype	phenotype
9 A— B—	9 M
3 aa B—	3 L′
3 A— bb	3 L
1 aa bb	1 M

If L and L' cannot be distinguished at the level of the phenotype, the ratio is 10:6.

3. $B \rightarrow A \rightarrow C \rightarrow D$

4. two

5. The end product of the wild-type gene contained at least two identical polypeptides. The mutant product of one allele interacted with the mutant product of the second allele in such a way that normal or nearly normal function resulted.

CHAPTER

13

DNA Function

Important Terms and Concepts

RNA is a single-stranded nucleic acid composed of a phosphate group, a ribose sugar, and one of four bases: adenine, guanine, cytosine, and uracil.

Transcription results in an RNA strand complementary to a DNA strand. Only one DNA strand is usually transcribed off a given region of a DNA helix. Along the helix, however, the strand being transcribed can switch numerous times, depending on the location of **promoters**. A promoter is the site for the initiation of transcription. For protein-encoding genes, the product of transcription is **mRNA**. Both **tRNA** and **rRNA** are transcribed, but not translated.

Translation is the ribosome-mediated production of a polypeptide, using the information encoded in mRNA. A **codon** is a sequence of three DNA bases that specifies an amino acid. The **genetic code** is the set of correspondences between nucleotide triplet codons and amino acids. The code is **degenerate**.

Wobble is the sloppy pairing between the anticodon of a tRNA molecule and the codon in the mRNA.

RNA processing consists of the addition and/or removal of one or more bases in an RNA molecule plus, in eukaryotes, the splicing out of **intron** transcript sequences. An intron is a sequence within the amino acid encoding region of a gene that is transcribed but not translated. An **exon** is the coding region in eukaryotic genes that is transcribed and translated into an amino-acid sequence.

Be sure that you have thoroughly read the entire chapter before you attempt any of the problems.

Solutions to Problems

1. Because RNA can hybridize to both strands, the RNA must be transcribed from both strands. This does not mean, however, that both strands are used as a template within each gene. The expectation is that only one strand is used within a gene but that different genes are transcribed in different directions. The most direct test would be to purify a specific RNA, and then hybridize it to the lambda genome. Only one strand should hybridize to the purified RNA.

2. **a.** The data do not indicate whether one or both strands are used for transcription in either case.

b. If the RNA is double-stranded, the percentage of purines (A + G) would equal the percentage of pyrimidines (U + C), and the AG/UC ratio would be 1.0. This is clearly not the case for *E. coli*, which has a ratio of 0.80. Therefore, *E. coli* RNA is single-stranded. The ratio for *B. subtilis* is 1.02. Either the RNA is double-stranded, or there is an equal number of purines and pyrimidines in each strand.

3. A single nucleotide change should result in three adjacent amino-acid changes in a protein. One and two adjacent amino-acid changes would be expected to be much rarer than the three changes. This is directly opposite of what is observed in proteins.

4. It suggests very little evolutionary change between *E. coli* and humans with regard to the translational apparatus. The code is universal, the ribosomes are interchangeable, the tRNAs are interchangeable, and the enzymes involved are interchangeable.

5. There are three codons for isoleucine: 5′AUU3′, 5′AUC3′, and 5′AUA3′. An unacceptable anticodon is 3′UAU5′ because it would also read 5′AUG3′—*met*. Possible anticodons (using Table 12-5) are 3′UAA5′ (complementary), 3′UAG5′ (complementary), and 3′AUI5′ (wobble).

6. a. Using Figure 12-19, there are eight cases in which knowing the first two nucleotides does not tell you the specific amino acid.

b. If you knew the amino acid, you would not know the first two nucleotides in the cases of Arg, Ser, and Leu (Figure 12-19).

7. The codon for amber is UAG. Listed below are the amino acids that would have been needed to be inserted to continue the wild-type chain and their codons:

glutamine	CAA, CAG*
lysine	AAA, AAG*
glutamic acid	GAA, GAG*
tyrosine	UAU*, UAC*
tryptophan	UGG*
serine	AGU, AGC, UCU, UCC, UCA, UCG*

In each case, the codon that has an asterisk by it would require a single base change to become UAG.

8. a. The codons for phenylalanine are UUU and UUC. Only the UUU codon can exist with randomly positioned A and U. Therefore, the chance of UUU is (1/2)(1/2)(1/2) = 1/8.

b. The codons for isoleucine are AUU, AUC, and AUA. AUC cannot exist. The probability of AUU is 1/8 and the probability of AUA is 1/8. The total probability is thus 1/4.

c. The codons for leucine are UUA, UUG, CUU, CUC, CUA, and CUG, of which only UUA can exist. It has a probability of 1/8.

d. The codons for tyrosine are UAU and UAC, of which only UAU can exist. It has a probability of 1/8.

9. a. 1U:5C: The probability of a U is 1/6, and the probability of a C is 5/6.

Codons	Amino acid	Probability	
UUU	phe	(1/6)(1/6)(1/6) = 0.005	phe = 0.027
UUC	phe	(1/6)(1/6)(5/6) = 0.023	
CCC	pro	(5/6)(5/6)(5/6) = 0.578	pro = 0.693
CCU	pro	(5/6)(5/6)(1/6) = 0.116	
UCC	ser	(1/6)(5/6)(5/6) = 0.116	ser = 0.139
UCU	ser	(1/6)(5/6)(1/6) = 0.023	

CUC	leu	(5/6) (1/6) (5/6) = 0.116	leu = 0.139
CUU	leu	(5/6) (1/6) (1/6) = 0.023	

1 Phe : 25 Pro : 5 Ser : 5 Leu

b. Using the same method as above, the final answer is 4 stop : 80 Phe : 40 Leu : 24 Ile : 24 Ser : 20 Tyr : 6 Pro : 6 Thr : 5 Asn : 5 His : 1 Lys : 1 Gin.

c. All amino acids are found in the proportions seen in Figure 12-19.

10. a. $(GAU)_n$ codes for Asp (GAU), Met (AUG), and stop (UGA). $(GUA)_n$ codes for Val (GUA), Ser (AGU), and stop (UAG). One reading frame contains a stop codon.

b. Each of the three reading frames contains a stop codon.

c. The way to approach this problem is to focus on one amino acid at a time initially. For instance, line 4 indicates that the codon for Arg can be AGA or GAG. Line 7 indicates it can be AAG, AGA, or GAA. Therefore, Arg is at least AGA. That also means that Glu is GAG (line 4). Lys and Glu can be AAG or GAA (line 7). Because no other combinations except the ones already mentioned result in either Lys or Glu, no further decision can be made with respect to them. However, taking wobble into consideration, Glu may also be GAA, which leaves Lys as AAG.

Next, focus on lines 1 and 5. Ser and Leu can be UCU and CUC. Ser, Leu, and Phe can be UUC, UCU, and CUU. Phe is not UCU, which is seen in both lines. From line 14, CUU is Leu. Therefore, UUC is Phe, and UCU is Ser.

The footnote says that line 13 is in the correct order. In line 13, if UCU is Ser (see above), then Ile is AUC, Tyr is UAU, and Leu is CUA.

Continued application of this approach will allow the assignment of an amino acid to each codon.

11. *Mutant 1*: A simple substitution of Arg for Glu exists, suggesting a nucleotide change. Two codons for Arg are AGA and AGG, and one codon for Ser is AGU. The final U for Ser could have been replaced by either an A or a G.

Mutant 2: The Trp codon (UGG) changed to a stop codon (UGA or UAG).

Mutant 3: Two frame-shift mutations occurred:

5′GCN CCN (-U)GGA GUG AAA AA(+U or C) UGU/C CAU/C3′.

Mutant 4: An inversion occurred after Trp and before Cys. The DNA original sequence was

3´CGN GGN ACC TCA CTT TTT ACA/G GTA/G5´

Therefore, the complementary RNA sequence was

5´GCN CCN UGG AGU GAA AAA UGU/C CAU/C3´

The DNA inverted sequence became

3´CGN GGN ACC AAA AAG TGA ACA/G GTA/G5´
^ ^

Therefore, the complementary RNA sequence was

5´GCN CCN UGG UUU UUC ACU UGU/C CAU/C3´
^ ^

12. old: AAA/G ĀGU CCA UCA CUU AAU GCN GCN AAA/G

new: AAA/G GUC CAU CAC UUA AUG GCN GCN AAA/G
+

Plus (+) and minus (-) are indicated in the appropriate strands.

13. e. With an insertion, the reading frame is disrupted. This will result in a drastically altered protein from the insertion to the end of the protein (which may be much shorter or longer than wild-type because of altered stop signals).

14. If the anticodon on a tRNA molecule also was four bases long, with the fourth base on the 5' side of the anticodon, it would suppress the insertion. Alterations in the ribosome can also induce frame shifting.

15.

3′	CGT	ACC	ACT	GCT	5′
5′	GCA	TGG	TGA	CGA	3′
5′	GCA	UGG	UGA	CGU	3′
3′	CGU	ACC	ACU	GCA	5′
	Ala	Trp	stop	nothing	

16. f, d, j, e, c, i, b, h, a, g

17. a. (40 aa) (3 nucleotides/aa) = 120 nucleotides

b. This is actually a three-point cross. At meiosis, the alignment is

mutant 1:	Gln	C^{I}	A^{II}	A/G
mutant 2:	Ser	A	G	U/C

The progeny from this cross are the two parental types plus

Arg	AGA/G	1 CO in II
Arg	CGN	1 CO in I
His	CAC/U	1 CO in II
Asn	AAC/U	DCO
Lys	AAA/G	1 CO in I

Notice that the distance between any two adjacent nucleotides is constant. This means that the frequency of recombination between adjacent nucleotides should be constant. The data on Asn can be ignored because they represent a DCO. Likewise, the data on Arg can be ignored because the Arg frequency results from two separate crossovers. The best data are twice the frequency of His or Lys, which is 4×10^{-7}.

18. Cells in long-established culture lines usually are not fully diploid. For reasons that are currently unknown, adaptation to culture frequently results in both karyotypic and gene dosage changes. This can result in hemizygosity for some genes, which allows for the expression of previously hidden recessive alleles.

19.a-b. First look for stop signs. Next look for the initiating codon, AUG (TAC in DNA). Only the upper strand contains a code exactly five amino acids long:

DNA 3'	TAC	GAT	CTT	TAA	GGC	ACT 5'
RNA 5'	AUG	CUA	GAA	AUU	CCG	UGA 3'
protein	Met	Leu	Glu	Ile	Pro	stop

The strand, obviously, is read from right to left as written in your text and is written above in reverse order from your text.

c. Remember that polarity must be taken into account. The inversion is

<--------

DNA 5′ TAC ATG CTA GAA ATT GCC TGA ATT GAT CAT GTA 3′

^

start

RNA stop 3′CUU UAA CGG ACU UAA CUA GUA 5′

amino acids HO-7 6 5 4 3 2 1-NH_3

d. 5′ UAC AUG AUC AUU UCA CGG AAU UUC UAG 3′

^ ^

start stop

Codon 4 is 5′ UCA 3′, which codes for ser.

Self-Test

1. If the anticodon sequence is 3′AUG 5′, what is the sequence of the nontranscribed DNA?

2. A mutation in the β-globin gene, a component of hemoglobin, results in a shortened polypeptide even though the immature (before splicing) mRNA transcript was of the correct length. Propose two different causes for the short polypeptide.

3. What would be the effect of an inversion within a gene?

4. What would be the effect of an inversion of an entire gene, including all its control sequences?

5. What is the role of tRNA?

6. What is the role of rRNA?

7. Are tRNA and rRNA transcribed? Translated?

8. Distinguish between missense and nonsense mutations.

9. What is the source of rRNA at the chromosomal level?

10. The DNA of a species contains 33 percent G. What can you say about the percentage of G seen in the RNA?

Solutions to Self-Test

1. 3'CAT5'

2. One cause could be the introduction of a stop codon. A second cause could be a change of the splicing site so that splicing eliminates more bases in the mRNA than it should.

3. At the very least, the amino acid sequence would be changed for the inverted region. Stop codons might also be generated.

4. No effect should be seen.

5. A tRNA molecule acts as an adapter between mRNA and amino acids.

6. Although rRNA molecules are found in ribosomes, their exact function is unknown.

7. Both are transcribed, neither is translated.

8. A missense mutation changes a codon so that it stands for a wrong amino acid. A nonsense mutation changes a codon so that it means "stop" to the translation system.

9. The nucleolar organizing region (NOR), located at the secondary constriction, contains the rRNA genes. It is the site of ribosome assembly.

10. The percentage of G in DNA is not a good predictor of the percentage of G in the RNA. Differential transcription of genes occurs. Furthermore, the species may have a higly repetitive, G-C rich fraction, which would not be transcribed.

CHAPTER

14

The Structure and Function of Chromosomes in Eukaryotes

Important Terms and Concepts

A **eukaryotic chromosome** is composed of a single, continuous molecule of DNA complexed with proteins. One class of proteins associated with the DNA is the **histones**. There are four major histones: **H2A, H2B, H3**, and **H4**. Two molecules of each of the histones form a structure called a **nucleosome**. About 140 base pairs of DNA wrap twice around a nucleosome. About 60 base pairs of DNA, called the **linker DNA**, extend between nucleosomes. Nucleosomes associate to form the **solenoid structure.**

There is a lack of correlation between the amount of DNA in a species and the complexity of the species. This is explained in part by the existence of **repetitive DNA sequences, intervening sequences**, and **spacer DNA**, none of which code for proteins. In *Drosophila* **polytene chromosomes**, each **chromomere** corresponds to a gene.

Genes that are derived from a common ancestral gene constitute a **gene family.** They are often clustered together, along with **pseudogenes.** Pseudogenes are nonfunctional copies of genes; they are nonfunctional because of evolutionary changes.

DNA complexed with protein is called **chromatin. Euchromatin,** which is usually light staining, contains genetically active genes. **Heterochromatin** which is densely staining, is thought to be nonfunctional. **Constitutive heterochromatin** are portions of the genome that are always heterochromatic. Frequently, the DNA contained in this region is **satellite DNA.** Constitutive heterochromatin may play a structural role in the chromosome. **Facultative heterochromatin** is euchromatic in some cells and heterochromatic in other cells. It consists of genes that are nonfunctional in the specific cell in which it is seen. One example is the **inactivated X chromosome** constituting the **Barr body.**

The **nucleolus** forms in the **nucleolar organizing region** (NO or NOR), which is located at the **secondary constriction** of chromosomes. It is the site of rRNA synthesis and ribosome assembly.

A **position effect** is seen when genes are translocated from euchromatin into or near heterochromatin.

Normal development and function require a balance of genetic functioning. When that balance is disturbed, through either an excess or a lack of genetic material, abnormalities result. In heterogametic species, there must be a mechanism to compensate for the differing amounts of genetic material between the sexes. This compensation is known as **dosage compensation.** The mammalian solution is **X-chromosome inactivation.** The presence of an inactivated X can result in **variegation** of phenotype.

Sex determination is achieved in different ways in different species.

Be sure that you have thoroughly read the entire chapter before you attempt any of the problems.

Solutions to Problems

1. c.

2. Triploids result in unpaired chromosomes being randomly distributed to the daughter cells during the reductional division. However, the equational divisions result in equal amounts of material going to the two poles.

a. A conventional sequence would show unequal amounts of chromosome material segregating at the first division. During the second division, equal amounts of chromosomal material would segregate.

b. A reversed sequence of divisions would show equal segregation at the first division and unequal amounts of chromosome material going to the two poles at the second division.

3. Homologous pairing of chromosomes is not understood. Several authors have suggested that chromosomes are nonrandomly positioned in the nucleus and that the random distribution seen in C-mitoses is an artifact of chromosome preparation. Colcemid and colchicine destroy the mitotic apparatus that aligns the chromosomes.

These authors suggest that chromosomes are anchored by their telomeres to the interphase nuclear membrane and that control over maintenance of position is passed back and forth between the nuclear membrane and the mitotic apparatus. In this view, homologous pairing as it is currently recognized is simply a closer association of homologs that are continuously located next to each other. This closer association is brought about by the synaptonemal complex, which is composed of protein, and perhaps by other proteins that have yet to be recognized. This closer pairing is clearly directed by sequence in some fashion; otherwise, the pairing of inversion heterozygotes cannot be explained.

4. The most viable explanation in the literature is that metaphase I does not occur until a minimal percentage of the genome is synapsed. Because inversions might require a longer period for synapsis in the inverted region, the remainder of the genome has a longer time in which to experience crossing-over. The most direct way to test this would be to observe chromosomes during prophase I and compare time of synapsis for normal and inversion lines.

5. Mutation is random; paramutation is directed and involves a specific interaction between specific alleles. A number of different hypotheses could be proposed to explain the observations.

6. The region of the heterochromatic knob contains a recognition point for whatever mechanism of migration that is in operation here.

7. The heterochromatin is likely to consist of repeated sequences. If so, it would renature far faster than the rest of the DNA. The nucleoli-like structures can be isolated by centrifugation. The heterochromatic DNA and the DNA from the supernumary chromosomes may differ in base sequence enough to allow isolation by centrifugation. If satellites exist that contain the DNA in question, the DNA can be purified by pouring it through a column to which denatured DNA from cells that do not contain it are attached. The single-stranded DNA obtained can then be cloned. The cloned DNA can be sequenced and used to do in situ hybridization.

8. Sequences present in the testes are not present in the salivary gland cells. This suggests that the process of polytenization does not include all DNA sequences. The missing sequences are GC-rich satellite sequences.

9. a. The following factors will affect the resolving power of the technique:

1. The specific activity (amount of radioactivity per amount of RNA) of the RNA.

2. The number of copies of DNA in each chromosome.

3. The degree of ploidy per cell.

4. The degree of denaturation of the DNA.

5. The type of radioactivity involved (determines both frequency of decay and size of the spot on the film).

6. The length of time of film exposure.

7. The size of the silver grain on the emulsion.

b. Excess cold RNA will compete with the radioactive RNA if it has the same sequence, but excess cold RNA will not compete with the radioactive RNA if it has a different sequence.

c. You infer that the RNA is complementary to the DNA. This leads to the further inference that the region to which binding occurred is the region from which the RNA is transcribed.

10. Some of the suggestions that have been made are

1. They are responsible for the homologous pairing seen in all *Drosophila* cells.

2. They facilitate synapsis.

3. They are part of the centromeres, with an unknown function.

11. Purify ribosomal sequences on a CsCl gradient (assuming that they are heavier than main band) and in situ hybridize the labeled sequences to chromosome preparations. This will localize the genes.

The rRNA genes can also be cloned, restriction-mapped, and located in the DNA sequences by Southern blotting. This will reveal copy number, arrangement of spacers, clustering, similarity of spacers, etc. The DNA also can be sequenced.

12. The mutants are most likely deletions of rRNA and tRNA gene clusters.

Crosses between the different, independently assorting mutants can be performed. For example, the cross $M—1/M—1^+ \; M—2^+/M—2^+ \times M—1^+/M—1^+ \; M—2/M—2^+$ will yield 1/4 double heterozygote progeny ($M—1/M—1^+ \; M—2/M—2^+$). If these mutants affect the same function, a 2:1 ratio may be observed (lethality). A 3:1 ratio would suggest that different functions are affected.

Isolate rDNA from wild-type, label it, and in situ hybridize it to polytene chromosomes from a specific minute and a strain that complements it. The same could be done with the tRNAs.

Self-Test

1. What is the relationship between genes and the DNA wrapped around the nucleosome?

2. If two *Xenopus* frogs each heterozygous for the anucleate deletion are crossed, what would be the result?

3. In cats, an X-linked gene controls coat color. One allele, X^b, results in black, another allele, X^o, results in orange.

a. What is the coat color of a heterozygous female?

b. If a male cat has the same coat color as a heterozygous female, what is his chromosomal constitution?

4. If a chromosomally normal woman is heterozygous for hemophilia, an X-linked recessive disorder, could she ever have hemophilia?

5. Pseudogenes are never transcribed. What kinds of evolutionary changes could result in nontranscription of an originally active gene?

Solutions to Self-Test

1. The average protein-encoding gene is approximately 100 amino acids long. That requires 3 × 100 DNA bases, at a minimum. Genes have associated promoter regions and they may have intervening sequences (introns). In addition, immature mRNA transcripts have leaders and trailers, also coded by a gene. Thus, the average gene may be several thousand base pairs long. The DNA associated with a nucleosome is, on average, 200 base pairs. Simple mathematics should convince you that one gene cannot wrap around one nucleosome. There does not seem to be any correlation between a gene and a nucleosome. The nucleosome must be viewed as a simple packaging mechanism.

2. The cross is $Aa \times Aa$. The *AA* progeny should have two nucleoli, the *Aa* progeny should have one nucleolus, and the *aa* progeny should have no nucleoli. The *aa* progeny would be expected to die as soon as the maternally derived ribosomes stored in the egg are used up during development.

3. a. Because of X-inactivation, females would be a mixture of black and orange patches.

b. Males with the same coat color must have two X chromosomes. Therefore, they are XXY.

4. Because of X-inactivation, it is theoretically possible for a heterozygous woman to have hemophilia.

5. Promoter sequences may have been deleted or mutated so that initiation of transcription does not occur. A mutated control factor (protein) may bind to the initiation site permanently and block initiation.

CHAPTER

15

Manipulation of DNA

Important Terms and Concepts

DNA is **denatured** by heat through the disruption of hydrogen bonds. The **melting point** is the temperature at which one-half of the DNA is no longer hydrogen-bonded. The resulting single-stranded DNA will undergo **reassociation**, or **reannealing**, when the temperature is decreased. **Hybridization** can be conducted through the use of two different DNAs. G—C rich DNA is more stable at a given temperature than A—C rich DNA.

In situ hybridization allows the localization of DNA sequences coding for an RNA sequence.

The genome can be fractionated on the basis of rate of reannealing or by centrifugation in CsCl. Specific sequences can sometimes be isolated using ethidium bromide with centrifugation in CsCl, RNA polymerase binding followed by enzyme digest, the use of reverse transcriptase, the use of restriction enzymes, and many other techniques.

Host restriction is the phenomenon whereby a given phage is capable of normal infection in one or several bacterial species. It is due to the action of **restriction nucleases** that degrade any DNA not specifically modified for protection from the nucleases of a given host cell. The host DNA is **modified** for protection. **Restriction enzymes** cleave DNA from any source at a specific site, unless the DNA has been modified.

The relative order of restriction-enzyme cleavage sites can be determined to provide a **restriction map.**

Recombinant-DNA technology involves a number of different techniques: **cloning, generation of sticky ends, tailing, blunt-end ligation,** the use of **vectors** such as **plasmids** and several phages, **shotgunning**, the formation of a **gene bank** or **gene library, Southern blotting,** the use of reverse transcriptase to generate **cDNA (complementary DNA)**, **chromosome walking, sequence determination, site-directed mutagenesis,** and **gene synthesis.**

Sequence determination has led to the discovery of **overlapping genes** in some viruses.

Transgenic organisms are organisms that develop from a cell into which new DNA has been introduced. Both **gene inactivation** and **regulation** have been studied in this way. **Gene therapy** has been conducted in animals.

Amniocentesis involves the removal and analysis of fetal amniotic fluid and cells. A number of genetic diseases can be diagnosed this way.

Restriction fragment length polymorphism (RFLP) has been used in the diagnosis of genetic disease.

Be sure that you have thoroughly read the entire chapter before you attempt any of the problems.

Solutions to Problems

1. AAGCTT occurs, on average, every 4^6 bases. CCGG occurs, on average, every 4^4 bases.

2. The lower the G—C content, the lower the temperature at which heat denaturing will occur. This is because it takes more energy to break the three hydrogen bonds in a G—C pair than it does to break the two hydrogen bonds of an A—T pair. The DNA with a 0.88 (G + C)/(A + T) ratio will melt at the lowest temperature. The DNA with the 1.2 ratio will melt at the highest temperature.

3. Both restriction enzymes recognize a six-base sequence, so both would be expected to have approximately the same number of recognition sites per genome. The major difference between the two is that EcoRI leaves staggered ends while SmaI leaves blunt ends. Staggered ends are much easier to manipulate during cloning because of the base specificity inherent in them.

4. Isolate the DNA and separate the two strands. Test each strand for the ability to hybridize with mRNA produced by the phage.

5.

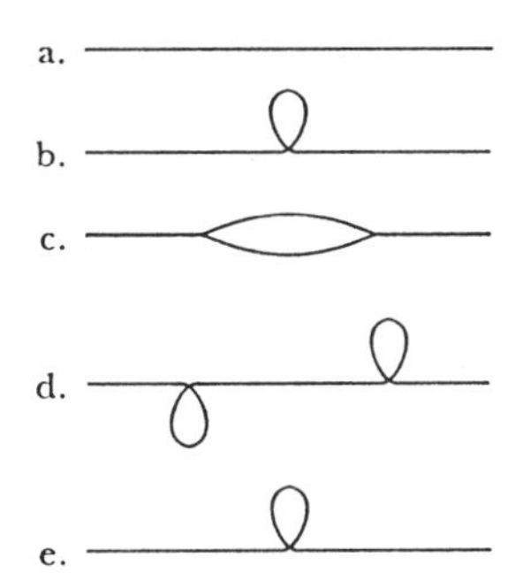

6. Any reasonable answer will do here. You might isolate protein from mutant and wild-type and do comparative electrophoresis. If you see a difference, you will have identified the gene product but not the function of the product.

7. a. In situ hybridization would localize the dAT to nucleus or cytoplasm.

b. In situ hybridization would also localize the dAT to chromosomal regions.

c. Denature crab DNA and hybridize it with poly-dAT. Separate single-stranded from double-stranded DNA on a hydroxyapatite column. Examine the double-stranded DNA by electron microscopy. If poly-dAT is linked to other DNA sequences, long single-stranded tails will be observed.

Alternatively, you could use rapid renaturation to enrich for repetitive sequences and then either clone it or do a base composition analysis to see if it is poly-dAt.

8. The structures result from palindromic DNA.

9. On the average, redundant sequences of 0.3 kb in length are interspersed every 0.8 kb in the genome.

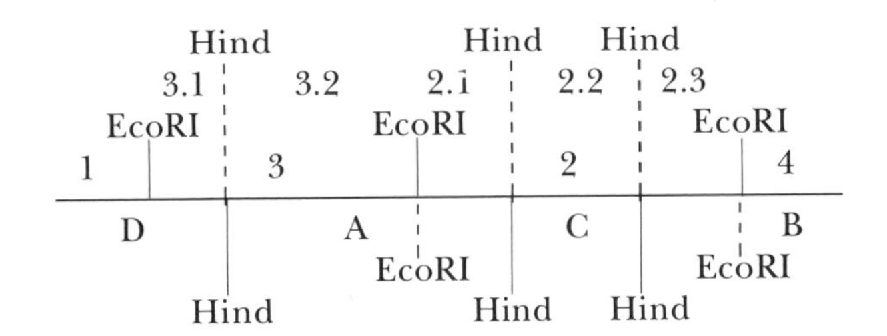

11. a. In situ hybridization using the polytene chromosomes would be the easiest way to identify chromosomal regions. Use whole-cell DNA from embryos lacking the region of interest. Next, using the clones that do not hybridize with the whole-cell DNA, do in situ hybridization onto wild-type polytene chromosomes.

b. Either Southern blotting or colony hybridization would identify a clone coding for a specific tRNA.

12. Reading from the bottom the sequence is

Hind-Hae-Hae-Hind-Hae-Hae-Hae-Hind-Hind-Hae-Hind-Hae-EcoRI

13. Reading from the bottom up,

left column: GGTACAACTATATATCAATTATAAAC

right column: GGATCTATTCTTATGATTATATAG

14. The tRNA sequence can be used to identify its DNA source through Southern blotting. Combined with restriction mapping, this will determine whether there are spacers between the genes. Sequencing the DNA will also yield the gene arrangement and information regarding spacers.

If nuclear RNA transcripts are isolated, they can be hybridized to the previously defined DNA. This will allow for a determination of whether the spacers are transcribed.

Mutations are difficult to study when there are multiple copies of the same gene. Assuming that a mutation in a tRNA gene results in no transcription (a deletion, a point mutation that prevents transcription or introduces a nonsense codon within the gene), mutant production of the tRNA should be reduced as compared with wild-type. Deletions also can be detected through Southern blotting.

15. Serine would be substituted for glycine in the D gene (missense mutation). In the E gene, the AUG start codon would be eliminated and the E gene would not be translated.

16. The XY male contains every sequence found in the XX female, plus the DNA sequence of the Y chromosome. Therefore, the DNA from a female can be used to purify DNA from the Y chromosome. Isolate, denature, and fix the DNA from a female to a column. Pour denatured DNA from a male through the column several times to remove all sequences complementary to the female. The remaining single-strand DNA will be Y chromosome DNA.

17. This problem assumes a random distribution of nucleotides.

AluI $(1/4)^4$ = every 256 nucleotide pairs

EcoRI $(1/4)^6$ = every 4096 nucleotide pairs

AcyI $(1/4)^4(1/2)^2$ = every 1024 nucleotide pairs

18. a.

1: *A1, B2*

2: *A2, B1*

3: *A1, B1*

4: *A2, B2*

b. Spores 3 and 4 are parental types, occurring in 70 percent of the population. Spores 1 and 2 are recombinants, occurring in 30 percent of the population.

c. *A* ______30______ *B*

19. a. The restriction map of pBR322 with the mouse fragment inserted is shown below. The 2.5-kb and 3.5-kb fragments would hybridize to the pBr322 probe.

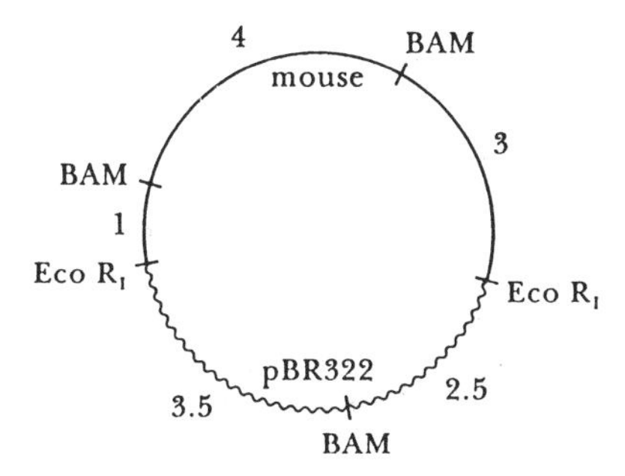

b. A protein 400 amino acids long requires a minimum of 1200 nucleotide bases. Only fragment 3 is long enough (3000 bp) to contain two or more copies of the gene.

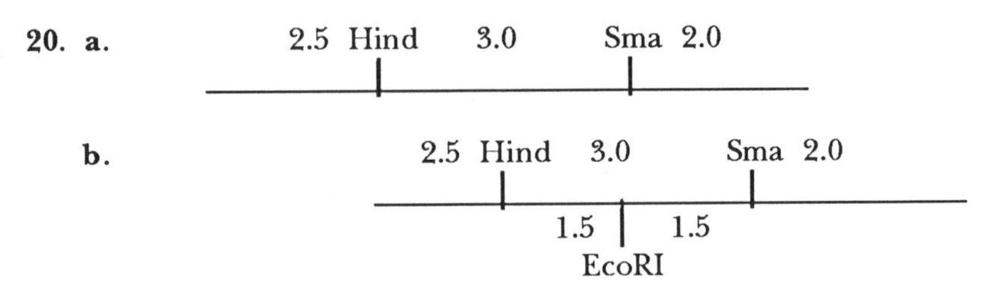

21. a. 1200 nucleotides

b. At 2 hours the viral transcript contains a nontranscribed intron sequence, which does not hybridize to the cDNA. The RNase removed the sequence, leaving behind 500 and 700 base fragments.

500 700

By 10 hours, the transcript has been spliced out, and a perfect hybrid forms between the 1200 base viral mRNA and the cDNA.

1200

c. It takes a minimum of 2 hours to transcribe and splice the mRNA and then translate it into protein.

22. If the yeast plasmid remains unintegrated, then it replicates independently of the chromosomes. During meiosis, the daughter plasmids would be distributed to the daughter cells, resulting in 100 percent transformation. This was observed in type C.

If one copy of the plasmid is inserted, when crossed with a *leu2*$^-$ line, the resulting offspring would have a ratio of 1 *leu*$^+$:1 *leu*$^+$. This is seen in type A and type B.

When the resulting *leu*$^+$ cells are crossed with standard *leu2*$^+$ lines, the data from type A cells suggest that the inserted gene is segregating independently of the standard *leu2*$^+$ gene, and data from type B cells suggest that the inserted gene is located in the same locus as the standard *leu2*$^+$ allele. Therefore, the gene in type A cells did not insert at the *leu2* site, and the gene in type B cells inserted at the *leu2* site.

23. Assuming that the same genes from different species have approximately the same base sequence, use the ß-tubulin gene cloned from *Neurospora* to isolate the ß-tubulin gene from *Podospora*.

Isolate plasmids from *Neurospora* using ethidium bromide and cesium chloride.

Cleave the *Podospora* DNA with EcoRI. Next, use denatured DNA from the cloned *Neurospora* gene to isolate the sequence from single-stranded *Podospora* DNA. Finally, melt the double-stranded DNA, releasing the desired sequence.

Once the desired sequence has been isolated, cleave the *E. coli* plasmid with EcoRI. Mix this DNA with the isolated sequence and add ligase.

Some of the plasmids will contain the desired sequence; they will lack resistance to tetracycline. The rest of the plasmids will not have the desired sequence; they will be resistant to tetracycline.

Introduce the plasmids into *E. coli* cells that are kan^S tet^S using a calcium chloride precipitation. Select for kan^R, killing off all *E. coli* that did not incorporate the plasmid.

Identify which resulting clones contain the desired sequence by isolating DNA from each clone, transferring it to a nitrocellulose filter, denaturing it, and probing with P^{32}-labeled DNA from *Neurospora.*

24.

EcoRI 8.0 | 7.4 | 4.5 | 2.9 | 6.2

BamHI 6.0 | 12.9 | 10.1

6.0 2.0 7.4 3.5 1.0 2.9 6.2

(X = 3.5)

25. a.

	AA	*BB*	*CC*	*DD*	*AB*	*AC*	*AD*	*BC*	*BD*	*CD*
17.5	—	—			—	—	—	—	—	
15.0			—	—		—	—	—	—	—
8.4	—	—	—	—	—	—	—	—	—	—
4.9		—		—	—		—	—	—	—
3.7	—		—		—	—	—	—		—
2.3	—	—	—	—	—	—	—	—	—	—
1.2	—		—		—	—	—	—		—

AD and *BC* are identical. The rest are all different.

b. The differences in restriction sites come from differences in DNA sequence. There is no evidence on which to base a judgment of either trivial or potentially adaptive differences.

c. The sequence that gave rise to the G8 probe is located on chromosome 4.

d. For each family, construct a 2 × 2 table for each polymorphism. Do not include people who marry into the family. Calculate X^2. This is done below for the relevant polymorphism in each family.

Venezuela

	Disease	No disease	Total
C present	19	1	20
C *absent*	*0*	*16*	*16*
Total	19	17	36

$$X^2 = \frac{(19-10.6)^2}{10.6} + \frac{(1-9.4)^2}{9.4} + \frac{(0-8.4)^2}{8.4} + \frac{(16-7.6)^2}{7.6}$$
$$= 6.7 + 7.5 + 8.4 + 9.3 = 31.9, \text{ highly significant}$$

United States

	Disease	No disease	Total
A present	13	7	20
A absent	0	1	1
Total	13	8	21

$$X^2 = \frac{(13-12.4)^2}{12.4} + \frac{(7-7.6)^2}{7.6} + \frac{(0-12.3)^2}{12.3} + \frac{(1-0.4)^2}{0.4}$$

$$= 0.029 + 0.047 + 12.3 + 0.9 = 13.276, \text{ highly significant}$$

Huntington's disease is linked with *C* in the family from Venezuela and with *A* in the family from the United States.

e. In each case, test for the relevant polymorphism by digesting with HindIII and probing with G8.

In the family from Venezuela, there is one crossover individual (VI, 5) among the 20 that carry the *C* polymorphism. Therefore, the G8 probe is (100%)(1)/20 = 5 m.u. from the Huntington's disease gene. If the person tests positive for the *C* polymorphism, there is a 95 percent chance of having the gene for Huntington's disease and a 5 percent chance of not having it. If the person tests negative for the *C* polymorphism, there is a 5 percent chance that he has the Huntington's disease gene and a 95 percent chance that he does not have the gene.

The situation in the family from the United States is unclear in comparison to the situation in the family from Venezuela. In the U.S.

family, the *A* polymorphism is present in three people with no family history of Huntington's disease who married into the family in question. The lower X^2 value in this family, as compared to the one from Venezuela, is a reflection of these individuals. Their presence makes it impossible to identify crossover individuals unambiguously. It can be assumed that the G8 probe is identifying a polymorphism that is approximately 5 m.u. from the Huntington locus, just as it did in the family from Venezuela. However, the conclusions from testing of the individual in question would vary with the polymorphism genotype of his affected parent.

For instance, if the affected parent were *AA* and the individual in question were *A*—, the chance of his having inherited the Huntington's disease gene would be 50 percent. If, however, the affected parent were *AD* and the individual in question were *A*—, the chance of his having inherited the Huntington's disease gene would be 95 percent, unless the unaffected parent also carried *A*. In that case, his risk would be 50 percent.

f. The G8 probe can be used to identify the region in which the Huntington's disease gene is located. The locus can be isolated by means of chromosome walking. The gene can be transcribed and translated, and the protein product can be identified.

g. Once the protein product of the Huntington's disease gene is identified, members of other families can be tested for the protein directly.

h. One exceptional person was identified already: Venezuela VI, 5, who is a crossover between the polymorphism and the Huntington's disease gene. In the U.S. pedigree there is no individual who is an obligate crossover.

26. a. All individuals have the 5-kb band, indicating that the band sequence is not involved with the gene in question. All affected individuals, except the last son, have the 3-kb and 2-kb bands. These two bands are not seen in unaffected individuals. The suggestion is that the two bands are close to or part of the gene in question.

b. The last affected son indicates that the two bands are not part of the gene in question. He represents a crossover between the gene in question and the two bands.

c. The 2-kb and 3-kb bands are closely linked to the dominant allele. Their presence in an individual would indicate a high risk of developing the disorder, while their absence would indicate a low risk of developing the disorder. Exact risk cannot be stated until the map units between the two bands and the gene in question are determined.

Although a rough estimate of map units can be made from the pedigree, the sample size is too small to make the estimate reliable.

Self-Test

1. RFLP DNA analysis is now being used in paternity suits. The following data were generated for one such case. Is the alleged father possibly the true father?

Mother	Child	Alleged father
———	———	
———		———
	———	———
———	———	
———	———	———
———	———	———

2. RFLP analysis is also now being used in criminal cases such as rape and murder. What complications exist here that do not exist in paternity suits?

3. When a fetus is being diagnosed for sickle-cell anemia and the mother is heterozygous, would contamination of the amniotic fluid by maternal cells lead to a false conclusion?

4. A linear fragment of DNA is cleaved with the individual restriction enzymes A and B, and then with a combination of the two enzymes. The fragments obtained are

enzyme A	3.0 kb, 5.6 kb
enzyme B	2.5 kb, 6.1 kb
enzymes A and B	0.5 kb, 2.5 kb, 5.6 kb

Draw a restriction map.

5. The DNA from three closely related species is heat-denatured. The melting points are A = 86.1°, B = 86.3°, and C = 85.7°. Which two species are closest? Why?

Solutions to Self-Test

1. The child does not have any band that could not have come from either his mother or the alleged father. Furthermore, he has a band present in the alleged father but not in his mother. It is possible that the alleged father is the true father.

2. There are two major complications: degradation of DNA and contamination of DNA. DNA degrades very rapidly under certain conditions. The environmental effects (heat, light, humidity, etc.) on DNA have not yet been determined. DNA from someone other than the perpetrator (from sloughed cells, shed hair, coughing, etc.) of a crime may be ubiquitous in the area that the crime occurred. This could lead to erroneous conclusions.

3. Yes. The maternal cells would indicate normal hemoglobin in the fetus.

4.

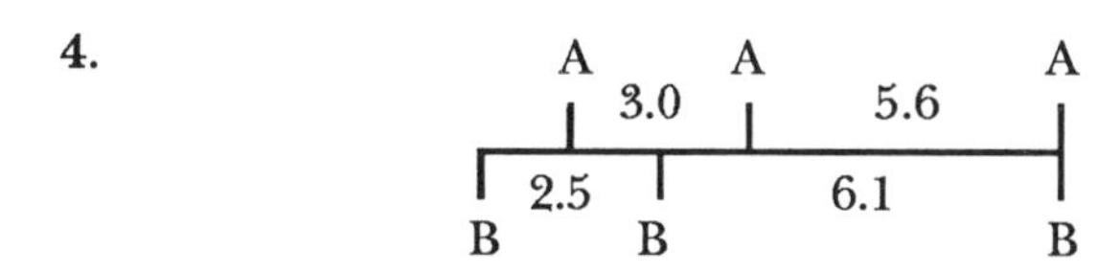

5. Species A and B have the least difference between melting points, which means the least difference between G-C : A-T ratio. Because the DNA total content is closest between these two, the best interpretation is that they are the closest species. However, other supporting data would have to be gathered before this conclusion would be anything but tentative.

CHAPTER

16

Control of Gene Expression

Important Terms and Concepts

Gene regulation is the regulation of transcription for specific genes. In prokaryotes, it is most often mediated by proteins that react to environmental signals by raising or lowering rate of transcription.

An **operon** consists of two or more cistrons plus the regulatory signals that affect transcription. Regulation occurs through the **promoter**, the ***I* locus**, and the **operator**. The **coordinately controlled genes** are transcribed in a **polycistronic** message. **Polar mutations** affect the gene within which they map and also reduce or eliminate the expression of all genes distal to the site of mutation in the polycistronic message.

Negative control is the blocking of transcription by a repressor protein that binds to an **operator**. The relief of **repression** is called **induction**.

The ***lac* operon** is under negative control. In the *lac* operon, **constitutive mutants** result in continuous expression in an unregulated fashion. Some are mutants of the *I* locus, I^-. The *I* locus determines the synthesis of a **repressor molecule**, which blocks activation of a gene or genes. An **operator constitutive mutation, O^c**, also results in unrepressed synthesis.

Positive control is the activation of transcription by a protein factor. **Catabolic repression** of the *lac* operon is an example of positive control. Here, the operon is activated by the presence of a large amount of CAP-cAMP.

Genes involved in the same metabolic pathway are frequently tightly clustered on prokaryotic chromosomes.

Feedback inhibition is the inhibition of the first enzyme in a metabolic pathway by the end product of that pathway, as exemplified in tryptophan biosynthesis. This is achieved by a process known as **attenuation**, an alteration of the secondary structure of the newly formed mRNA in the leader region.

Eukaryotic regulation is controlled by the **promoter, enhancers,** and **upstream activating sequences.** Several proteins have been identified that interact with these sites. Some steroid hormones also bind at these sites.

Genetic redundancy and **gene amplification** are cellular mechanisms that ensure an adequate supply of vital gene products.

Be sure that you have thoroughly read the entire chapter before you attempt any of the problems.

Solution to Problems

1. The *I* gene determines the synthesis of a repressor molecule, which blocks expression of the *lac* operon and which is inactivated by the inducer. The presence of the repressor, I^+, will be dominant to the absence of a repressor, I^-. I^s mutants are unresponsive to an inducer. For this reason, the gene product cannot be stopped from interacting with the operator and blocking the *lac* operon. Therefore, I^s is dominant to the I^+.

2. O^c mutants do not bind the repressor product of the *I* gene, and therefore, the *lac* operon associated with the O^c operator cannot be turned off. Because an operator controls only the genes on the same DNA strand, it is cis (on the same strand) and dominant (cannot be turned off).

3. **a.** Comparing lines 1 and 2, *a* and *b* in a + or – state do not affect the expression of the *Z* gene. Therefore, *b* is the *Z* gene.

In line 6, the *I* gene is functioning in a trans configuration, indicating that *c* is *I*. This leaves *a* as the *O* region, which is confirmed by line 7.

b. *Line 1*: $a^- = I^c$ or O^c

Line 2: $c^- = I^c$ or O^c

Line 3: $c^- = I^c$ or I^s

Line 4: $a^- = O^c$, and $c^- = I^c$ or I^s

Line 5: $a^- = O^c$ or O^o, and $c^- = I^c$

Line 6: $a^- = O^c$ or O^o, and $c^- = I^c$

Line 7: $a^- = O^c$, and $c^- = I^c$ or I^s

4.

	ß-Galactosidase		Permease	
Part	No lactose	Lactose	No lactose	Lactose
a	+	+	−	+
b	+	+	−	−
c	−	−	−	−
d	−	−	−	−
e	+	+	+	+
f	+	+	−	−
g	−	+	−	+

5. a. A lack of only E_1 or only E_2 function indicates that both genes have enzyme products that are responsible for a conversion reaction. Because the two genes are in different linkage groups, they cannot be regulated by a single operator and promoter like the *Z* and *Y* genes of the *lac* operon. Type 3 mutants must be mutants of a site that produces a diffusible regulator of the E_1 and E_2 genes. The type 3 mutants identify a site that produces either a repressor (like *I* in the *lac* operon) or an activator (analogous to CAP) of the other two genes.

b. Separate operator and promoter mutants might be found for each gene.

6. If there is an operon governing both genes, then a frame-shift mutation could cause the stop codon separating the two genes to be read as a sense codon. Therefore, the second gene product will be incorrect for almost all amino acids. However, there are no known polycistronic messages in eukaryotes. The alternative, and better, explanation is that both enzymatic functions are performed by the same gene product. Here, a frame-shift mutation beyond the first function, carbamyl phosphate synthetase, will result in the second half of the protein molecule being nonfunctional.

7. Nonpolar Z^- mutants cannot convert lactose to allolactose, and thus, the operon is never induced.

8. Because very small amounts of the repressor are made, the system as a whole is quite responsive to changes in lactose concentration. In the heterodiploids, repressor tetramers may form by association of polypeptides encoded by I^- and I^+. The operator binding site binds two subunits at a time. Therefore, the repressors produced may reduce operator binding, which in turn would result in some expression of the *lac* genes in the absence of lactose.

9. An operon is turned off by the mediator in negative control, and the mediator must be removed for transcription to occur. An operon is turned on by the mediator in positive control, and the mediator must be added for transcription to occur.

10. The *lacY* gene produces a permease that transports lactose into the cell. A *lacY* $^-$ gene could not transport lactose into the cell, so ß-galactosidase will not be induced.

11. Activation of gene expression by trans-acting factors occurs in both prokaryotes and eukaryotes. In both cases, the trans-acting factors interact with specific sequences that control expression of cis genes.

In prokaryotes, proteins bind to a specific DNA sequence, which is regulated by the binding protein and which, in turn, regulates one or more downstream cistrons.

In eukaryotes, highly conserved sequences such as CCAAT and enhancers increase transcription controlled by the downstream TATA box promoter. Several proteins have been found to bind to the CCAAT sequence, upstream GC boxes, and the TATA sequence in *Drosophila*, yeast, and other organisms. Specifically, the Sp1 protein recognizes the upstream GC boxes of the SV40 promoter and many other genes, GCN4 and GAL4 proteins recognize upstream sequences in yeast, and many hormones bind to specific sites on the DNA (e.g., estrogen binding to a sequence upstream of the ovalbumin gene in chicken oviduct cells). Additionally, the structure of some of these trans-acting DNA-binding proteins is quite similar to the structure of binding proteins seen in prokaryotes. Further, protein-protein interactions are important in both prokaryotes and eukaryotes. For the above reasons, eukaryotic regulation is now thought to be very close to the model for regulation of the bacterial *ara* operon.

12. The bacterial operon consists of a promoter region that extends approximately 35 bases upstream of the site where transcription is initiated. Within this region is the promoter. Inducers and repressors, both of which are trans-acting proteins that bind to the promoter region, regulate transcription of associated cistrons in cis only.

The eukaryotic cistron has the same basic organization. However, the promoter region is somewhat larger. Also, enhancers up to several thousand nucleotides upstream or downstream can influence the rate of transcription. A major difference is that eukaryotes have not been demonstrated to have polycistronic messages.

Self-Test

1. Using the standard format for the *lac* gene, identify which products are made and whether they are induced or made constitutively.

   a. $I^s P^+ O^+ Z^+ Y^+$

   b. $I^s P^+ O^+ Z^+ Y^+$

   c. $I^s P^+ O^C Z^+ Y^+$

   d. $I^+ P^- O^+ Z^+ Y^+$

   e. $I^+ P^+ O^C Z^+ Y^+$

2. What is the mechanism of attenuation?

3. Can cells operate in a positive feedback loop?

4. It has been suggested that intron transcripts have the potential to control transcription of a gene through negative feedback. How could this occur?

5. What is the difference between positive control and positive feedback?

Solutions to Self-Test

1.

	ß-galactosidase		Permease	
	No lactose	Lactose	No lactose	Lactose
a.	–	–	–	–
b.	–	–	–	–
c.	+	+	+	+
d.	–	–	–	–
e.	+	+	+	+

2. In the tryptophan operon, transcription pauses in the leader sequence. During this period, the ribosome that is translating the transcript (remember transcription is coupled to translation) proceeds through the leader to a given point in the presence of excess tryptophan. This forces the leader to assume a secondary configuration that causes termination of transcription. With a lack of excess tryptophan, the ribosome is halted at an earlier point, the leader assumes another secondary structure as a consequence, and transcription continues. This is a negative feedback control.

3. A positive feedback loop means that the end product of a metabolic pathway would induce the formation of more of itself. This is a

prescription for disaster in any system because it leads to uncontrolled behavior.

4. The intron transcripts would have to be complementary to the promoter region of the transcribed strand. By pairing with this region, they could block transcription.

5. Positive control is the initiation of transcription of a gene by a protein factor transcribed off a different gene. Positive feedback is the initiation of transcription by the end product of transcription from the same gene.

CHAPTER

17

Mechanisms of Genetic Change I: Gene Mutation

Important Terms and Concepts

Mutations consist of a change in DNA sequence. **Spontaneous** mutations occur at a low rate in all cells; the mechanism for spontaneous mutations is unknown. **Induced** mutations are caused by one or more environmental agents.

A **tautomeric shift** results in a **transition** mutation: a purine is substituted for a purine or a pyrimidine is substituted for a pyrimidine. **Transversion** mutations substitute a purine for a pyrimidine or a pyrimidine for a purine.

Frame-shift mutations result in a change in the reading frame; they are caused by slipped mispairing during the replication of repeated sequences. **Deletion** mutations can be caused by replication errors or recombinational errors.

Depurination, the loss of the nitrogenous purine base from the nucleotide, can result in a mutation. **Deamination** of cytosine yields uracil, which pairs with adenine (a transition).

Base analogs induce mutation at a high rate; they are incorporated into the DNA and alter base pairing. **Alkylating agents** chemically modify the nitrogenous bases to cause mutations. **Intercalating agents** insert themselves between the nitrogenous bases, causing additions and deletions.

DNA is **repaired** by several different systems: the **SOS system**, **detoxification** by **superoxide dismutase**, the **photoreactivating enzyme**, **alkyltransferase**, **excision repair**, the **AP endonucleases**, the **DNA glycosylase repair pathway**, the **mismatch repair system**, and **recombinational repair**.

The **Ames test** is a screening test for possible mutagens.

Be sure that you have thoroughly read the entire chapter before you attempt any of the problems.

Solutions to Problems

1. **a.** A transition mutation is the substitution of a purine for a purine or the substitution of a pyrimidine for a pyrimidine. A transversion mutation is the substitution of a purine for a pyrimidine, or vice versa.

b. Both are base-pair substitutions. A silent mutation is one that does not alter the function of the protein product from the gene, because the new codon codes for the same amino acid as did the nonmutant codon. A neutral mutation results in a different amino acid that is functionally equivalent, and the mutation therefore has no adaptive significance.

c. A missense mutation results in a different amino acid in the protein product of the gene. A nonsense mutation causes premature termination of translation, resulting in a shortened protein.

d. Frame-shift mutations arise from the addition or deletion of a base, which alters the reading frame for translation and, therefore, the amino-acid sequence from the site of the mutation to the end of the protein product of the gene. Frame-shift mutations can result in nonsense (stop) mutations.

2. A nonsense mutation results in a stop codon. Often the truncated protein does not eliminate enzyme activity or change the structure of a structural protein to the point where it is nonfunctional. A frame-shift

mutation usually results in many amino-acid substitutions or, quite frequently, the introduction of a stop codon. When many amino acids in a protein are changed, the enzyme activity usually is eliminated, and any structural protein would be highly altered.

3. The Streisinger model proposed that frame-shifts arise when loops in single-stranded regions are stabilized by slipped mispairing of repeated sequences. In the *lac* gene of *E. coli*, a four-base-pair sequence is repeated three times in tandem, and this is the site of a hot spot.

The sequence is 5′ CTGG CTGG CTGG CTGG 3′. During replication the DNA must become single-stranded in short stretches for replication to occur. As the new strand is synthesized and becomes hydrogen-bonded to the template strand, it can pair out of register with that strand by a total of four bases. Depending on which strand, new or template, loops with regard to the other, there will be an addition or deletion of four bases, as diagramed below:

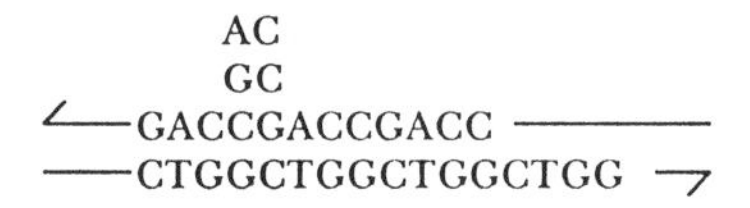

In this diagram, the upper strand looped as replication was occurring. The loop is stabilized by base-pairing on either strand. As replication continues at the 3′ end, an additional copy of GACC will be synthesized, leading to an addition of four bases. This will result in a frame-shift mutation.

4. a. In Problem 17-3, had the lower strand looped, the result would have been a deletion in the newly synthesized upper strand.

b. As with the bar-eye allele in *Drosophila*, misalignment of homologous chromosomes during recombination results in one chromosome with a duplication and the other with a deletion:

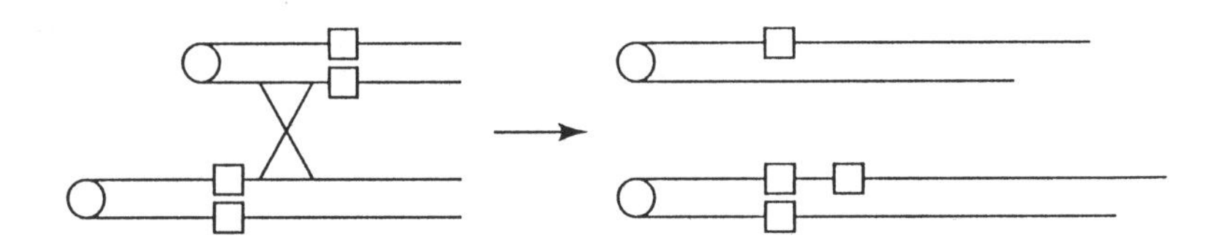

The above models are supported by DNA sequencing results.

5. a. Depurination results in the loss of adenine or guanine from the nucleotide. This apurinic site cannot specify a complementary base, which blocks replication. Under certain conditions, replication proceeds with a random insertion of a base opposite the apurinic site. In three-fourths of these insertions, a mutation will result.

b. Deamination of cytosine yields uracil. If left unrepaired, adenine is paired with uracil during replication, resulting in a transition mutation.

6. 5-bromouracil is an analog of thymine. It undergoes tautomeric shifts at a higher frequency than thymine and, therefore, is more likely to pair with G than is thymine during replication. At the next replication this will lead to a GC pair rather than the original AT pair.

Ethylmethanesulfonate is an alkylating agent that produces 0-6-ethyl-guanine. This will pair with thymine, which leads from a GC pair to an AT pair at the next replication.

7. An AP site is an apurinic or apyrimidinic site. AP endonucleases introduce chain breaks by cleaving the phosphodiester bonds at the AP sites. Some exonuclease activity follows, so that a number of bases are removed. The resulting gap is filled by DNA pol I and then sealed by DNA ligase.

When UV damage occurs in *E. coli,* several compounds may bind to the damaged site, resulting in the so-called bulky adducts. The *uvrA*, *uvrB*, and *uvrC* gene products recognize a distortion in the DNA helix. Again, excision is followed by gap-filling and ligation.

8. a. Mismatch repair occurs when a mismatched nucleotide is inserted during replication. The new, incorrect base is removed and a proper base is inserted. The enzymes involved can distinguish between new and old strands because, in *E. coli,* the old strand is methylated.

b. Recombination repair occurs when AP sites and UV photodimers block replication (there is a gap in the complementary strand). Recombination occurs with one strand from the sister DNA molecule, which is normal in both strands. This produces two DNA molecules, each with a lesion in one strand. Each lesion is then repaired normally.

c. Proofreading by DNA polymerase will result in the removal of an incorrect base and the insertion of a correct base.

9. Leaky mutants are mutants with an altered protein product that retains a low level of function. Enzyme activity may, for instance, be reduced rather than abolished.

10. The wild-type contained a gene that increased the spontaneous mutation rate. This new gene seems to be unlinked to *ad—3*. Call the new gene *B*. Cross *A* (*ad—3* B^+) × wild-type (*ad—*3^+ *B*). The progeny should reflect independent assortment.

Progeny: 1/4 *ad—3 B*

1/4 *ad—3 B*$^+$

1/4 *ad—3*$^+$ *B*

1/4 *ad—3*$^+$ *B*$^+$

Further crosses should verify the above.

11. a. Because 5´UAA3´ does not contain G or C, a transition to a GC pair in the DNA cannot result in 5´UAA3´. 5'UGA3´ and 5´UAG3´ have the DNA antisense-strand sequence of 3´ACT5´ and 3´ATC5´. A transition to either of these stop codons occurs from the nonmutant 3´ATT5´. A DNA sequence of 3´ATT5´ results in an RNA sequence of UAA, itself a stop codon.

b. Yes. An example would be 3´UGG5´, which codes for *trp*, to 3'UAG5'.

c. No. In the three stop codons the only base that can be acted upon is G (in UAG, for instance). Replacing the G with an A would result in 3´UAA5´, a stop codon.

12.a. and **b.** *Mutant 1*: Most likely a deletion. It could be caused by radiation.

Mutant 2: Because proflavin causes either additions or deletions of bases and because spontaneous mutation can result in additions or deletions, the most probable cause was a frame-shift mutation by an intercalating agent.

Mutant 3: 5-BU causes transitions, which means that the original mutation was most likely a transition. Because HA causes GC to AT transitions and HA cannot revert it, the original must have been a GC to AT transition. It could have been caused by base analogs.

Mutant 4: The chemical agents cause transitions or frame-shift mutations. Because of spontaneous reversion only, the original mutation must have been a transversion. Nitrogen mustard could have caused the original mutation.

Mutant 5: HA causes transitions from GC to AT, as does BU. The original mutation was most likely an AT to GC transition, which could be caused by base analogs.

c. The suggestion is a second-site reversion linked to the original mutant by 20 map units (m.u.) (prototrophs equal one-half of the recombinants).

13. To understand these data, recall that half of the progeny should come from the wild-type parent.

a. A lack of revertants suggests either a deletion or an inversion within the gene.

b. *Prototroph A*: Because 100 percent of the progeny are prototrophic, a reversion at the original mutant site may have occurred.

Prototroph B: Half of the progeny are parental prototrophs, and the remaining prototrophs, 28 percent, are the result of the new mutation. Notice that 28 percent is approximately equal to the 22 percent auxotrophs. The suggestion is that an unlinked suppressor mutation occurred, yielding independent assortment with the *nic* mutant.

Prototroph C: There are 496 "revertant" prototrophs (the other 500 are parental prototrophs) and 4 auxotrophs. This suggests that a suppressor mutation occurred in a site very close (100% [4 × 2]/1000 = 0.8 m.u.) to the original mutation.

14. a. To select for a nerve mutation that blocks flying, place *Drosophila* at the bottom of a cage and place a poisoned food source at the top of the cage.

b. Make antibodies against flagellar protein and expose mutagenized cultures to the antibodies.

c. Do filtration through membranes with various-sized pores.

d. Screen visually.

e. Go to a large shopping mall and set up a rotating polarized disk. Ask the passersby to look through the disk for a free evaluation of their vision and their need for sunglasses. People with normal vision will see light with a constant intensity through the disk. Those with polarized vision will see alternating dark and light.

f. Set up a Y tube and have the flies or unicellular algae crawl to either light or dark.

g. Set up replica cultures and expose one of the two plates to low doses of UV.

Self-Test

1. Name two ways to increase the number of copies of a sequence.

2. Many repeated sequences diverge over time, yet some remain absolutely unchanged. The process by which spontaneous changes in repeated sequences are reversed is known as rectification. No one knows how rectification occurs. Propose two mechanisms.

3. If you wanted to introduce small deletions, how would you do it?

4. Which chemical group distinguishes new DNA strands from old? Which chemical group is frequently added to a base to avoid restriction endonuclease activity?

5. What would constitute a secondary cure for the recessive human disorder xeroderma pigmentosum?

6. Would you expect there to be a correlation between gene length and mutation rate?

Solutions to Self-Test

1. Mispairing during recombination and slipped pairing during replication.
2. Recombinational repair; sister-chromatid exchange, if it occurs by the process of recombination.
3. The easiest way would be through the use of intercalating agents.
4. Methyl groups in both cases.
5. Remaining out of the sun or using complete sun blockers.
6. The longer the gene, the larger the number of sites available for mutation.

CHAPTER

18

Mechanisms of Genetic Change II: Recombination

Important Terms and Concepts

Crossing-over occurs during prophase I of meiosis. It involves the **breakage and reunion** of DNA molecules. **Chiasmata** are the sites of crossing-over.

The **Holliday model** involves the creation of **heteroduplex** DNA, which can undergo **branch migration**. A number of enzymes are postulated to function in this process. The model accounts for **gene conversion**, **polarity**, and **co-conversion**.

Site-specific recombination occurs between two specific sequences that need not be homologous.

Be sure that you have thoroughly read the entire chapter before you attempt any of the problems.

Solutions to Problems

1. Gene conversion is a meiotic process of directed change in which one allele directs the conversion of a partner allele to its own form. Instead of meiosis ending with equal numbers of both alleles, gene conversion results in an excess of one allele and a deficiency of the other. In contrast, mutation is undirected change that can occur during both meiosis and mitosis.

2. Gene conversion may result in a deviation from a 4:4 ratio, with the order unimportant. The following asci show gene conversion: 3, 6.

Ascus 4 is also produced by gene conversion. To recognize it as such, recall the sequence that gives rise to the eight meiotic products in *Neurospora.* The pattern generated could be produced only if the two DNA strands have a region of mismatch.

3. In the first case, 1′ is being converted to 1′+. In the second case, 1″ is being converted to 1″+. The difference in frequency is due to polarity.

4. A fixed break point is the point at which a DNA strand breaks and begins unwinding as the first step in recombination. The highest level of gene conversion is seen at this point.

Gene conversion has occurred in cistrons 1, 2, and 3, and the conversions all are from mutant to wild-type. Therefore, most likely one piece of heteroduplex DNA extended across the three cistrons.

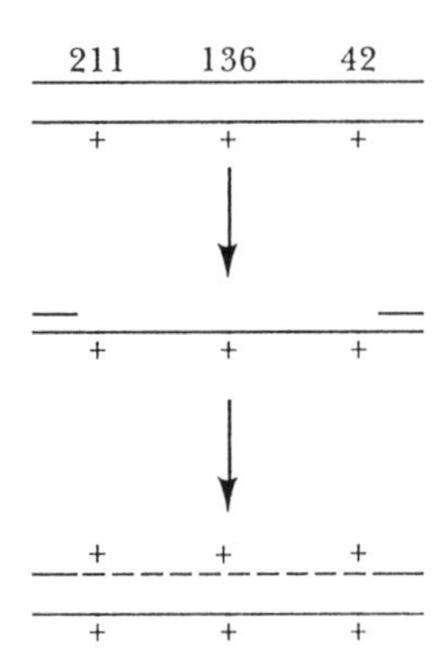

These data are compatible with the idea but do not prove that the heteroduplex DNA initiates from a site to the left of cistron 1, corresponding to a promoter of a polycistronic message.

5. The actual mechanism of sister-chromatid exchange induction by mutagens is unknown but would be expected to vary with mutagen effects.

The end point must be a break in a DNA strand, very likely as part of postreplication repair.

6. First notice that gene conversion is occurring. In the first cross, a_1 converts (1:3). In the second cross, a_3 converts. In the third cross, a_3 converts. Polarity is obviously involved. The results can be explained by the following map, where hybrid DNA enters only from the left.

$$a_3 \text{ --- } a_1 \text{ --- } a_2$$

7. Rewrite the cross and results so that it is clear what they are.

P	$A\ (m_1\ m_2\ m^+)\ B \times a\ (m^+\ m^+\ m_3)\ b$	
F_1	$A\ (m_1\ m_2\ m^+)\ B$	parental
	$A\ (m_1\ m^+\ m^+)\ b$	recombinant
	$a\ (m_1\ m_2\ m^+)\ B$	recombinant
	$a\ (m^+\ m^+\ m_3)\ b$	parental

Next, note the frequency of each allele:

$$m_1 : m^+ = 3:1$$

$$m_2 : m^+ = 1:1$$

$$m_3 : m^+ = 1:3$$

Two gene conversion events have occurred, involving m_1 and m_3.

To understand this at the molecular level, consider the following diagram:

A m_1 m_3^+ m_2 B		A m_1 m_3^+ m_2 B
A m_1 m_3^+ m_2 B		A m_1^+ m_3 m_2^+ b
a m_1^+ m_3 m_2^+ b	⟶	a m_1 m_3^+ m_2 B
a m_1^+ m_3 m_2^+ b		a m_1^+ m_3 m_2^+ b

	A m_1 m_3^+ m_2 B		A m_1 m_3^+ m_2 B
	A m_2^+ b		A m_1 m_3^+ m_2^+ b
⟶	a m_1 m_3^+ m_2 B	⟶	a m_1 m_3^+ m_2 B
	a m_1^+ m_3 m_2^+ b		a m_1^+ m_3 m_2^+ b

A single excision-repair event changed m^+ m_3 to m_1 m^+, and the other mismatches remained unrepaired.

8. The ratios for a_1 and a_2 are both 3:1. There is no evidence of polarity, which indicates that gene conversion as part of recombination is occurring. The best explanation is that two separate excision-repair events occurred and, in both cases, the repair retained the mutant rather than the wild-type.

9. **a** and **b**. A heteroduplex that contains an unequal number of bases in the two strands has a larger distortion than a simple mismatch. Therefore, the former would be more likely to be repaired. For such a case, both heteroduplex molecules are repaired (leading to 6:2 and 2:6) more often than one (leading to 5:3 or 3:5) or none (leading to 3:1:1:3). The preference in direction (i.e., adding a base rather than subtracting) is analogous to TT dimer repair. In TT dimer repair, the unpaired, bulged nucleotides are treated as correct and the strand with the TT dimer is excised.

A mismatch more often than not escapes repair, leading to a 3:1:1:3 ascus.

Transition mutations would not cause as large a distortion of the helix, and each strand of the heteroduplex should have an equal chance of repair. This would lead to 4:4 (two repairs each in the opposite direction), 5:3 (1 repair), 3:1:1:3 (no repairs or two repairs in opposite directions), and, less frequently, 6:2 (two repairs in the same direction).

c. Because excision repair excises the strand opposite the larger buckle (i.e., opposite the frame-shift mutation), the *cis* transition mutation will also be retained. The nearby genes are converted because of the length of the excision repair.

10. The easiest way to handle these data is to construct the following table, which shows the repair rates for 0, 1, and 2 hybrid DNA molecules.

	0.5 +/+	0.3 +/g_1	0.2 g_1/g_1
0.5 +/+	0.25	0.15	0.1
	(6:2)	(5:3)	(4:4)
0.3 +/g_1	0.15	0.09	0.06
	(5:3)	(3:1:1:3)	(3:5)
0.2 g_1/g_1	0.1	0.06	0.04
	(4:4)	(3:5)	(2:6)

The aberrant asci are all those that are not 4:4. The 4:4 asci occur 20 percent of the time. Correcting for them,

a. 6:2 = 25% / 0.8 = 31.25%

b. 2:6 = 4% / 0.8 = 5%

c. 3:1:1:3 = 9% / 0.8 = 11.25%

d. 5:3 = 30% / 0.8 = 37.5%

e. 3:5 = 12% / 0.8 = 15%

11. The map is

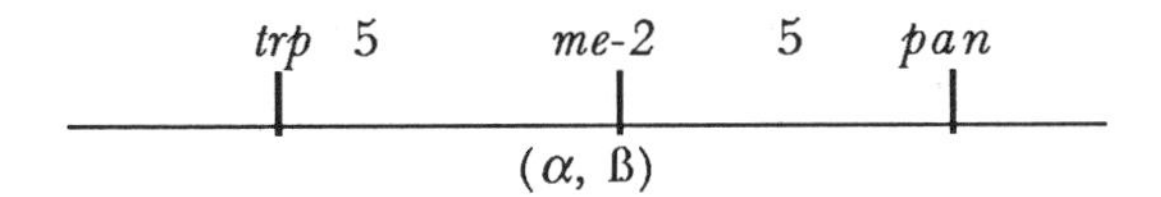

Rewrite the crosses and results to be sure that you understand them.

Cross 1: *trp* (α ß$^+$) *pan*$^+$ × *trp*$^+$ (α^+ ß) *pan*

Cross 2: *trp* (α^+ ß) *pan*$^+$ × *trp*$^+$ (α ß$^+$) pan

Note that all progeny must be α^+ ß$^+$, which requires a crossover between them, and that the order of α and ß are unknown.

Consider the first cross. If the sequence is *trp* α ß *pan,* then one crossover should lead to a high frequency of + + + +. The conventional double crossovers + + + - and - + + + should be equally frequent and of lower frequency than + + + +. The pattern - + + - would result from a triple crossover and would be least frequent. This is summarized in the tabulation below, along with the results if the opposite gene order is true.

Pattern	If *trp* α ß *pan*	If *trp* ß α *pan*	Frequency
+ + + +	1 CO	3 CO	56

− + + +	DCO	DCO	26
+ + + −	DCO	DCO	59
− + + −	3 CO	1 CO	16

For the second cross, the patterns and their interpretation are

Pattern	If *trp α β pan*	If *trp β α pan*	Frequency
+ + + +	3 CO	1 CO	15
− + + +	DCO	DCO	84
+ + + −	DCO	DCO	23
− + + −	1 CO	3 CO	87

Both crosses indicate that the sequence is *trp* α β *pan*, but these results are, on the surface, confusing. We see that a double-crossover event does not lead to reciprocal results and, in fact, one double-crossover product occurs as frequently as the single-crossover product. The difficulty is not in the cross but in thinking of the results in terms of a conventional Mendelian cross rather than in terms of gene conversion. Double crossovers are not occurring in the Mendelian sense. In both crosses, "crossing-over" between β and the *pan* allele is occurring at a much higher frequency than expected, which means that β is being converted at a higher level than is α. By convention, that means the polarity runs from *pan* toward *trp*. The asymmetry due to polarity is also seen in the *trp* + : + *pan* ratios in each cross.

12. Rewrite the original cross:

$$\text{P} \quad A\ x\ y^+ \times a\ x^+\ y$$

The progeny of parental genotypes will be like either of the two parents. The backcrosses are as follows, with prime indicating progeny generation.

Cross 1: $a'\ x^+\ y \times A\ x\ y^+ \rightarrow 10^{-5}$ prototrophs

Cross 2a: $A'\ x\ y^+ \times a\ x^+\ y \rightarrow 10^{-5}$ prototrophs

Cross 2b: $A'\ x\ y^+ \times a\ x^+\ y \rightarrow 10^{-2}$ prototrophs

Recombination is allowing for the higher rate of appearance of prototrophs. Cross 2 is obviously a backcross for some gene affecting the rate of recombination. Whatever that gene is, the allele in the *A* parent blocks recombination (cross 1 and cross 2a), and the allele in the *a* parent allows recombination (cross 2b). It is unlinked to the *his* gene since cross

2 yields results in a 1:1 ratio. The allele that blocks recombination (in *A*) is dominant, while the allele that allows recombination is recessive. This is demonstrated by the original cross, in which prototrophs occurred at the lower rate, and by cross 1.

To test this interpretation, one-fourth of the crosses between the $A\ x\ y^+$ and $a\ x^+\ y$ progeny should yield a high rate of recombination and therefore have a high frequency of prototrophs.

Self-Test

1. Can the Holliday model be applied to sister-chromatid exchange? If yes, what initial assumption must you make?

2. Some mutagens, in addition to causing sister-chromatid exchange, cause mitotic recombination between homologs. What does this suggest about the interphase location of homologous chromosomes?

3. Which of the following linear asci show gene conversion?

1	2	3	4	5	6
+	+	+	+	+	+
+	+	+	+	+	+
met	*met*	+	+	+	*met*
met	*met*	+	+	+	+
met	+	+	+	*met*	*met*
met	*met*	*met*	+	*met*	*met*
met	*met*	*met*	*met*	*met*	+
met	*met*	*met*	*met*	*met*	+

4. If branch migration occurs, what is the relationship between the initial break point and the chiasma that results? What do the data of Tease and Jones suggest about branch migration?

Solutions to Self-Test

1. The Holliday model can account for sister-chromatid exchange. It must be assumed that a break caused by a mutagen signals an enzyme with endonuclease function to cause a break in the homologous site on the sister chromatid.

2. It suggests that homologous chromosomes are located near each other in mitotic interphase. Although it is generally assumed that the distribution of chromosomes during interphase is random, there are several studies that indirectly indicate that chromosome distribution is nonrandom during interphase.

3. All asci except 5 show gene conversion.

4. There is not necessarily any relationship between an initial break point and the resulting chiasma. As branch migration proceeds, the center of the recombination intermediate can be at any point along the length of a chromosome. The probability is very large that resolution of the recombination intermediate will not occur at the exact site of the initial break point.

Because Tease and Jones found that the dark/light transition occurred right at the chiasma, the tentative suggestion is either that branch migration does not occur or that it occurs for a limited distance not detectable by light microscopy.

CHAPTER

19

Mechanisms of Genetic Change III: Transposable Genetic Elements

Important Terms and Concepts

Transposable genetic elements exist in both prokaryotes and eukaryotes. They move from location to location within the genome. Both the DNA sequence from which they are removed and the DNA sequence into which they are inserted are altered, potentially causing phenotypic changes that are scored as mutations.

Prokaryotic **insertion sequences**, termed **IS elements**, usually block the expression of all genes downstream in the operon from the site of insertion. These are **polar mutations**.

Prokaryotic **transposons** possess **inverted repeat (IR) sequences**, which are IS elements that flank one or more genes. If they are located on a plasmid, they can be passed during conjugation from organism to organism within a species or between closely related species. The transposon can move from plasmid to plasmid or between a plasmid and a bacterial chromosome. Transposition can be **replicative** or **conservative.**

There are several types of eukaryotic transposable genetic elements: the yeast **Ty elements**; the *Drosophila* **copia-like elements**, the **fold-back elements (FB)**, and the ***P* elements**; the maize **controlling elements.**

Retroviruses are RNA viruses that integrate into host chromosomes after a DNA copy of the genome is made using **reverse transcriptase.** When integrated, they are termed **proviruses.** They possess long terminal repeats like prokaryotic transposons.

In eukaryotes the transposable elements use an RNA intermediate, while in prokaryotes transposition occurs at the DNA level.

Be sure that you have thoroughly read the entire chapter before you attempt any of the problems.

Solutions to Problems

1. Hybridize *E. coli* containing the mutation with *E. coli* not containing the mutation and observe the hybrid molecules by electron microscopy.
Do a restriction analysis of wild-type and mutant DNA.

2. Polar mutations affect the transcription or translation of the part of the gene or operon on only one side of the mutant site, usually described as downstream. Examples are nonsense mutations, frame-shift mutations, and IS-induced mutations.

3. In replicative transposition a new copy of the transposable element is generated during transposition. The experiment by Ljungquist and Bukhari first demonstrated this through the use of restriction-enzyme digestion patterns.
In conservative transposition no replication occurs. The element is excised from its location and integrated into the new site. Kleckner and coworkers demonstrated this by constructing heteroduplexes and using them to infect cells.

4. R plasmids are the main carriers of drug resistance. They acquire these genes by transposition of drug-resistance genes located between IR (inverted repeat) sequences. Once in a plasmid, the transposon carrying drug resistance can be transferred upon conjugation if it stays in the R plasmid or it can insert into the host chromosome.

5. Boeke, Garfinkel, Fink, and their coworkers demonstrated that transposition of the Ty element in yeast involved an RNA intermediate. They constructed a plasmid using the Ty element. It had a promoter near the end of the element that could be activated by galactose and it had an intron inserted into the Ty transposon coding region. After transposition, they found that the new transposon lacked the intron sequence. Because intron splicing occurs only in RNA, there must have been an RNA intermediate.

6. *P* elements are transposons (genes flanked by inverted repeats, allowing for great mobility). Because they are transposons, they can insert into chromosomes. By inserting specific DNA between the inverted repeats of the *P* elements and injecting the altered transposons into cells, a high frequency of gene transfer will occur.

7. The $a_1 a_1$ *DtDt* plant has the *A* (target gene) inactivated by the insertion of a receptor element into it. The regulator is at the *Dt* locus. When crossed with a $a_1 a_1$ *dtdt* tester, the progeny should be $a_1 a_1$ *Dtdt*, colorless.

Excision of the receptor element restores the *A* gene function.

Black kernels would arise if the receptor element excises from the chromosome prior to fertilization. Dotted kernals arise from sporadic excision at a later stage in development of the kernel.

8. The best explanation is that the mutation is due to an insertion of a transposable element.

9. The sn^+ patches in an *sn* background and the occurrence of sn^+ progeny from an $sn \times sn$ mating mean that the sn^+ allele is appearing at a fairly high frequency. The *sn* allele is unstable, suggesting that an insertion element in the sn^+ gene results in *sn*.

10. a. The expression of the tumor is blocked in plant B. This suggests that either plant B can suppress the functioning of the plasmid that causes the tumor, or that plant A provides something to the tissue with the tumor-causing plasmid that plant B does not provide.

b. Tissue carrying the plasmid, when grafted to plant B, appears normal, but the graft produces tumor cells in synthetic medium. This indicates that the plasmid sequences are present and capable of functioning in the right environment. However, the production of normal type A plants from seeds from the graft suggests a permanent loss of the plasmid during meiosis.

11. *Cross 1:*

P $C/c^{Ds}\ Ac/Ac^+ \times c/c\ Ac^+/Ac^+$

F_1 1 $C/c\ Ac/Ac^+$ (solid pigment)

1 C/c Ac^+/Ac^+ (solid pigment)

1 c^{Ds}/c Ac/Ac^+ (unstable colorless or spotted)

1 c^{Ds}/c Ac^+/Ac^+ (colorless)

Cross 2:

P $C/c^{Ac} \times c/c$

F_1 1 C/c (solid pigment)

1 c/c^{Ac} (spotted)

Cross 3:

P C/c^{Ds} $Ac/Ac^+ \times C/c^{Ac}$ Ac^+/Ac^+

F_1 1 C/C Ac/Ac^+ (solid pigment)

1 C/c^{ac} Ac/Ac^+ (solid pigment)

1 C/C Ac^+/Ac^+ (solid pigment)

1 C/c^{ac} Ac^+/Ac^+ (solid pigment)

1 C/c^{Ds} Ac^+/Ac^+ (solid pigment)

1 C/c^{Ds} Ac^+/Ac (solid pigment)

1 c^{Ds}/c^{ac} Ac^+/Ac^+ (spotted)

1 c^{Ds}/c^{ac} Ac^+/Ac (spotted)

Self-Test

1. What evidence indicates that transposable elements in prokaryotes do not strictly jump from one location to another?

2. A retrovirus has 27 percent G. When it is in the provirus state, what will be its percentage of G?

3. Consider the following genetic units: episome, lysogenic virus, plasmid, retrovirus, and lytic virus. Arrange them on a continuum and justify the arrangement. The answer to this question requires a creative approach.

4. The virus causing AIDS is a retrovirus. Suggest a way that it could be detected in humans.

5. Some retroviral infections are associated with a high rate of malignancy. List several mechanisms that could account for this.

Solutions to Self-Test

1. If transposable elements did jump, there would be a correlation between loss at one location and gain at another. This is not observed. Also, intermediate cointegrate structures have been observed during transposition of some elements.

2. There is no way to predict the percentage of G in the DNA copy of the retrovirus, because a retrovirus is single-stranded and the percentage of C is not known.

3. Several different schemes could be proposed. One follows.

plasmid	an episome that has lost the ability to integrate
episome	a virus that has lost the ability to exist extracellularly
lytic virus	a virus that has lost the ability to undergo lysogeny
retrovirus	a virus that has lost the ability to cause lysis
lysogenic virus	a complete virus

4. Restriction analysis should be able to detect the AIDS virus.

5. If the virus specifically integrates into cells of the immune system, the functioning of the immune system is impaired. Thus, the immune system is unable to destroy the malignant cells, which are always being formed in multicellular organisms.

By integrating into specific genes that repair DNA lesions, the virus can eliminate repair functions. Thus, the genome can become more and more altered over time, leading to malignancy eventually.

The virus may itself carry genes that cause malignancy.

CHAPTER

20

The Extranuclear Genome

Important Terms and Concepts

Extranuclear genes exist in eukaryotes. These genes are located in organelles. They do not show Mendelian patterns of inheritance.

Organelles such as **chloroplasts** and **mitochondria** contain circular DNA. The phenotypes coded for by organelle DNA generally show a **maternal inheritance** pattern. **Segregation** is commonly seen among the progeny when two or more genetically different chloroplast or mitochondrial DNAs exist in the mother. **Recombination** and **extranuclear mutation** exist in organelle DNA. Organelle DNA can be mapped.

Most organelle-encoded polypeptides unite with nucleus-encoded polypeptides to form active proteins. These proteins function in the organelle.

Maternal inheritance does not always indicate extranuclear inheritance. For some characteristics, the maternal nuclear genome determines a progeny phenotype. This is called **maternal effect.**

Be sure that you have thoroughly read the entire chapter before you attempt any of the problems.

Solutions to Problems

1. Most organelle-encoded polypeptides unite with nucleus-encoded polypeptides to produce active proteins, and these active proteins function in the organelle.

2. Reciprocal crosses reveal cytoplasmic inheritance. Cytoplasmic inheritance also can be demonstrated by doing a series of backcrosses, using hybrid females in each case, so that the nuclear genes of one strain are functioning in cytoplasm from the second strain.

3. Maternal inheritance of chloroplasts results in the green/white color variegation observed in *Mirabilis.*

Cross 1: variegated female × green male → variegated progeny

Cross 2: green female × variegated male → green progeny

4. **a.** stop-start growth

   b. normal growth

5. Both yeast parents contribute mitochondria to the cytoplasm of the resulting diploid cell. Subsequent meiosis shows uniparental inheritance for mitochondria. Therefore, 4:0 and 0:4 asci will be seen.

6. **a.** The ant^R gene may be mitochondrial.

   b. Some of the petites must have been neutral petites, in which all mitochondrial DNA, including the ant^R gene, was lost. Other petites were suppressive, in which the ant^R gene was retained.

7. The genetic determinants of R and S are cytoplasmic and are showing paternal inheritance.

8. The cpDNA from *mt⁻ Chlamydomonas* is lost. The results of the crosses are

Cross 1: all morph 1; 2 kb and 3 kb bands

Cross 2: all morph 2; 3 kb and 5 kb bands

9. Both yeast parents contribute mitochondria to the cytoplasm of the resulting diploid cell. Subsequent meiosis shows uniparental inheritance for mitochondria. If no crossing-over occurs in the diploid fusion cell, asci will be of two types:

$$4\ oli^{R}\ cap^{R}$$

$$4\ oli^{S}\ cap^{S}$$

If crossing-over occurs, the recombinants will be of two types:

$$4\ oli^{R}\ cap^{S}$$

$$4\ oli^{S}\ cap^{R}$$

10. Both crosses show maternal inheritance of a chloroplast gene. The rare variegated phenotype is probably due to a minor male contribution to the zygote. Variegation must result from a mixture of normal and prazinizan chloroplasts.

11. If the mutation is in the chloroplast, reciprocal crosses will give different results, while if it is in the nucleus and dominant, reciprocal crosses will give the same results.

12. This pattern is observed when a maternal recessive nuclear gene determines phenotype. The crosses are

P   *dd* dwarf female × *DD* normal male

F_1 *Dd* dwarf (all dwarf because mother is *dd*)

F_2 3/4 *D*— : 1/4 *dd* normal (all normal because mother is *Dd*)

F_3 3/4 normal (mother is *D*—) : 1/4 dwarf (mother is *dd*)

13. After the initial hybridization, a series of backcrosses using pollen from B will result in the desired combination of cytoplasm A and nucleus B. With each cross the female contributes all of the cytoplasm and one-half the nuclear contents, while the male contributes one-half the nuclear contents.

14. a. Maternal inheritance is suggested.

b. "Why" can never be inferred. However, the net result was that a line carrying the nuclear genes of *E. hirsutum* contained the cytoplasm of *E. luteum*. This demonstrated that the very tall progeny were not the result of a hybrid nucleus.

15. Let male sterility be symbolized by MS.

a. A line that was homozygous for *Rf* and contained the male-sterility factor would result in fertile males. When this line was crossed, using females, with a line not carrying the restorer gene for two generations, the male-sterility trait would reappear.

b. The F_1 would carry the male-sterility factor and would be heterozygous for the *Rf* gene. Therefore, it would be fertile.

c. The cross is *Rf/rf* MS × *rf/rf* (no cytoplasmic transmission). The progeny would be 1/2 *Rf/rf* MS (fertile) and 1/2 *rf/rf* MS (sterile).

d. i. P *Rf-1/rf-1 Rf-2/rf-2* × *rf-1/rf-1 rf-2/rf-2* MS

F_1 1/4 *Rf-1/rf-1 Rf-2/rf-2* MS fertile

1/4 *Rf-1/rf-1 rf-2/rf-2* MS fertile

1/4 *rf-1/rf-1 Rf-2/rf-2* MS fertile

1/4 *rf-1/rf-1 rf-2/rf-2* MS male sterile

ii. P *Rf-1/Rf-1 rf-2/rf-2* × *rf-1/rf-1 rf-2/rf-2* MS

F_1 100% *Rf-1/rf-1 rf-2/rf-2* MS fertile

iii. P *Rf-1/rf-1 rf-2/rf-2* × *rf-1/rf-1 rf-2/rf-2* MS

F_1 1/2 *Rf-1/rf-1 rf-2/rf-2* MS fertile

1/2 *rf-1/rf-1 rf-2/rf-2* MS male sterile

iv. P *Rf-1/rf-1 Rf-2/Rf-2* × *rf-1/rf-1 rf-2/rf-2* MS

F_1 1/2 *Rf-1/rf-1 Rf-2/rf-2* MS fertile

1/2 *rf-1/rf-1 Rf-2/rf-2* MS fertile

16. The suggestion is that the mutants were cytoheterozygotes for streptomycin resistance.

17. Realize that the closer two genes are, the higher the rate of cosegregation. A rough map of the results is as follows.

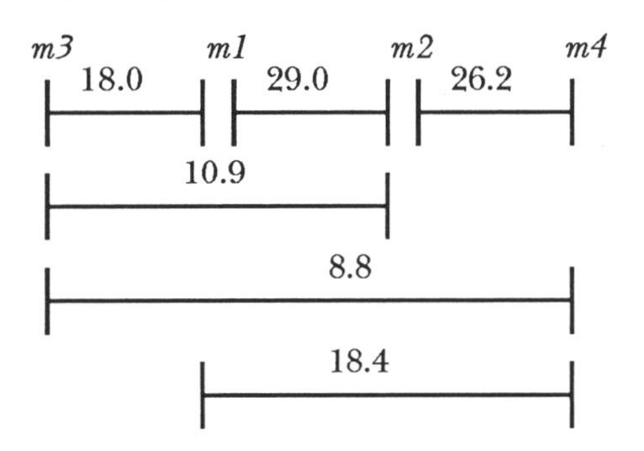

18. The red phenotype in the heterokaryon indicates that the red phenotype is caused by a cytoplasmic organelle allele.

19. a. Injection of normal mitochondria into wild-type. This should result in the wild-type phenotype in several generations unless there is an injection effect.

b. Injection of physiological saline solution into wild-type. This also should result in the wild-type phenotype in several generations unless there is an injection effect.

c. In order to be sure that a cytoplasmic factor rather than a nuclear gene is being transferred, all donor material should be derived from organisms with many alleles variant from the wild-type.

20. Notice that the results are not reciprocal, which indicates an extranuclear gene. Notice also that in the first cross, there is a 1:1 ratio of phenotypes, indicating a nuclear gene. This may be a case where the same function has a different location in different species. The two genes are incompatible in the hybrid. Let *cy* = cytoplasmic factor in *N. sitophila*, A^c = the nuclear allele in *N. crassa*, and A = the nuclear allele in *N. sitophila*. Aconidial is then A^c *cy*. The crosses are

$$A\ cy \times A^c \rightarrow 1/2\ A\ cy\ \text{(normal)} : 1/2\ A^c\ cy\ \text{(aconidial)}$$

$$A^c \times A\ \text{(no } cy \text{ contribution from male)} \rightarrow \text{all normal } (A \text{ and } A^c)$$

Neither parent was aconidial, because the sporeless phenotype requires the interaction of a nuclear allele from one species with a cytoplasmic factor from the other species.

21. Let poky be symbolized by (*c*). Let the nuclear suppressor of poky be symbolized by *n*. To do these problems, you cannot simply do the crosses in sequence. For instance, the parental genotypes in cross a must be written taking cross c into consideration.

	Cross	Progeny
a.	(+) + X (*c*) +	all (+) +
b.	(+) *n* X (*c*) +	1/2 (+) *c* : 1/2 (+) +
c.	(*c*) + X (+) +	all (*c*) +
d.	(*c*) + X (+) *n*	1/2 (*c*) + (= D) : 1/2 (*c*) *n* (= E)
e.	(*c*) *n* X (+) *n*	all (*c*) *n*
f.	(*c*) *n* X (+) +	1/2 (*c*) *n* : 1/2 (*C*) +

22. a. and **b.** Each meiosis shows uniparental inheritance, suggesting cytoplasmic inheritance.

c. Because *ant*^{r} is probably mitochondrial and because petites have been shown to result from deletions in the mitochondrial genome, *ant*^{r} may be lost in some petites.

23. a. The first tetrad shows a 2:2 pattern, indicating a nuclear gene, while the second tetrad shows a 0:4 pattern, indicating maternal inheritance and a cytoplasmic factor (mitochondrial).

b. The nuclear gene should always show a 1:1 segregation pattern. The mitochondrial gene could produce an ascus that was all *cyt2*$^+$.

c. Both produce proteins involved with the cytochromes. Either the products of the two genes affect different steps in mitochondrial function or they affect the same step if the two proteins interact to form one enzyme.

24. Consider the following hypothetical situation of three genes: —— *a* ——— *b* —— *c* ——— . If a deletion of one of them occurred, *b* by itself would be least likely to be the only gene deleted if the genes were closely linked. The closer *a* is to *b*, the less likely it is that *a* will be deleted if *b* is retained.

Applying the above to the data, the *apt-cob* pair had the lowest rate of loss (45 total), and the *apt—bar* pair had the highest rate of loss (207 total). This puts *cob* between *apt* and *bar*. The relative rate of loss is *apt—cob* : *cob-bar*, or 45 : 117 = 1 : 2.6. *apt* —— *cob* ——— *bar*

1 2.6

25. Remember that petites arise by deletion. Hybridization with any fragment means that the gene being tested is on that fragment.

Culture 1: *cap* and $rRNA_{large}$ are on the same fragment

Culture 2: $tRNA_4$ is not next to *cap, ery, oli,* or *par*

Culture 3: $tRNA_2$ and $tRNA_5$ are on the same fragment

Culture 4: *oli,* $rRNA_{large}$, and $rRNA_{small}$ are on the same fragment

Culture 5: *ery,* $rRNA_{large}$, and $rRNA_{small}$ are on the same fragment

Culture 6: *cap* and $tRNA_3$ are on the same fragment

Culture 7: *oli, par,* and $tRNA_2$ are on the same fragment

Culture 8: $rRNA_{small}$, $tRNA_4$, and $tRNA_5$ are not next to *cap, ery, oli,* or *par*

Culture 9: *ery* and $rRNA_{large}$ are on the same fragment

Culture 10: $tRNA_1$ and $tRNA_3$ are on the same fragment

Culture 11: *ery* and $rRNA_{large}$ are on the same fragment

Culture 12: *ery*, $rRNA_{large}$, and $rRNA_{small}$ are on the same fragment

Once you have identified each fragment, begin arranging them in overlapping order. For example,

Culture 1: *cap* $rRNA_{large}$

Culture 9: $rRNA_{large}$, *ery*

Culture 5: $rRNA_{large}$, *ery* $rRNA_{small}$

gives you *cap*—$rRNA_{large}$—*ery*—$rRNA_{small}$. The entire sequence is a circle:

cap—$rRNA_{large}$—*ery*—$rRNA_{small}$ -$tRNA_4$—$tRNA_5$—$tRNA_2$—*par*—*oli*—$tRNA_1$—$tRNA_3$—*cap*.

26. Some tetrads will show strain-1 type, some will show strain-2 type, and some will be recombinant.

27. a. No, during diploid budding all the progeny receive one type of mtDNA.

b. Most likely it is a plasmid or episome.

28. a. Use mitochondria from donors who differ on several nuclear genes from the recipient. Assess the recipient for those markers. If they were never transferred, this indicates that there was no accidental contamination of the mitochondria with nuclear DNA.

b. Inject some cells with the same volume of buffer but without mitochondria. Inject some cells with the same volume of buffer plus nucleoplasm. Inject some cells with the same volume of buffer plus cytoplasm lacking mitochondria. All of these controls will help to ensure that the observed effect was due to the mitochondria.

29. First, prepare a restriction map of the mtDNA using various restriction enzymes. Using the assumption of evolutionary conservation, Southern blot hybridize equivalent fragments from yeast or another organism in which the genes have already been identified.

30. a. The cytoplasm from senescent cultures is a mixture of normal and abnormal mitochondria. The mitochondrial types are distributed in different ratios to different spores. The abnormal mitochondria seem to have a replicative advantage over the normal since senescence seems, ultimately, to "win out" over normal nonsenescence. The rapidity of the onset of senescence seems to be related to the ratio of normal-to-abnormal mitochondria.

b. The mutation is an insertion of about 10 kb. It carries bands E and G and splits the original fragment into two fragments, B and C.

31. a. The plant is a mosaic: one cell line is normal and the other cell line has both cpDNAs. Assume that the plant is self-fertilizing.

Recall endosperm formation from Chapter 3. The endosperm is derived from two identical haploid female nuclei and one haploid male nucleus. In order for four bands to appear on the gel, lane 3, the cytoplasm of the meiocyte giving rise to the seed must have contained a mixture of both cpDNAs. This could occur only if a progenitor cell contained the original mutant and normal cpDNA, and segregation of the two types of chloroplasts did not occur.

In order to get homozygous cpDNA, seen in lanes 1 and 2, segregation of chloroplasts had to occur. Lane 1 is derived from normal cpDNA, which could have come from the normal cell line or the mutant line, through segregation. Lane 2 is derived from mutant cpDNA, through segregation.

b. Both *Gryllus* and *Drosophila* would be expected to have segregation of mitochondria, but it is not being seen in these rare females. Therefore, it must be hypothesized that their mitochondria all contain two genomes, one of which carries the mutation and one of which is normal. In other words, they are the products of a segregation of mitochondria that occurred in an earlier generation. In this case, all of their progeny would be expected to show the mixture seen in the rare females.

32. The two repeats, center to center, are separated by 83 kb. If the two direct repeats lined up as below and then experienced a recombination event, the two smaller circles would be 83 kb and 135 kb, each containing a single copy of the direct repeat.

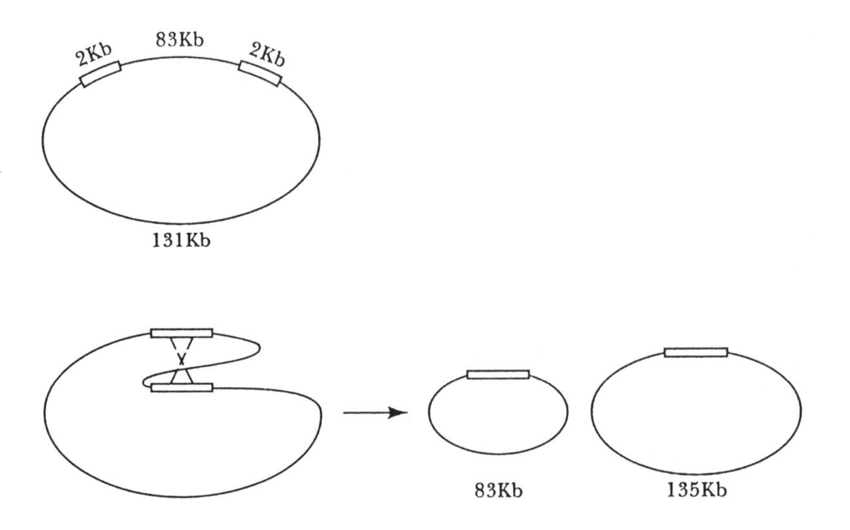

Self-Test

1. *Limnaea peregra* is a hermaphroditic snail that can reproduce either by crossing or by self-fertilization. Shell coiling in progeny is controlled by a maternal gene. A right-hand shell pattern is caused by the dominant gene s^+. The left-hand shell pattern is recessive. A homozygous right-hand shell is fertilized by a homozygous left-hand shell.

a. What is the phenotype of the offspring?

b. If the offspring self-fertilize, what are the genotypes and phenotypes in the F_2?

c. If the offspring self-fertilize, what are the genotypes and phenotypes in the F_3?

2. In the snail described in the previous problem, if a heterozygous left-handed shell is fertilized by a homozygous left-handed shell, what will be the phenotype of the progeny?

3. Heroin-addicted mothers give birth to heroin-addicted babies, whether or not the father is heroin-addicted. Nonaddicted mothers fertilized by addicted fathers give birth to nonaddicted babies. This maternal effect is not caused by a nuclear or a cytoplasmic gene. Suggest a mechanism for the maternal effect.

4. A strain of *Chlamydomonas* requires streptomycin in the culture medium for survival. Most strains are sensitive to streptomycin. How can you determine if the streptomycin-dependence is due to a cytoplasmic or a nuclear gene?

5. In mice, certain strains have a high rate of breast cancer, and other strains have a low rate of breast cancer. A cross between high females and low males results in approximately 90 percent breast cancer in females when they begin to nurse their offspring. The reciprocal cross does not result in breast cancer in the female offspring. However, if high males are crossed with low females and the female progeny are nursed by high females, approximately 90 percent of them develop breast cancer when they begin nursing their own offspring. How can you explain these results?

Solutions to Self-Test

1. **a.** The cross is

P $s^+s^+ \times ss$

F_1 s^+s (right-hand)

b.

F_1 $s^+s \times s^+s$

F_2 1 s^+s^+ : 2 s^+s : 1 ss (all right-hand)

c.

Cross 1:

F_2 $s^+s^+ \times$X s^+s^+

F_3 s^+s^+ (all right-hand)

Cross 2:

F_2 $s^+s \times s^+s$

F_3 1 s^+s^+ : 2 s^+s : 1 ss (all right-hand)

Cross 3:

F_2 $ss \times ss$

F_3 all ss (all left-hand)

2. all right-handed

3. The heroin in the mother's blood passes through the placenta, enters the baby's blood system, and causes addiction.

4. Cross a male-sensitive with a female-dependent. If the gene is cytoplasmic, nearly all of the progeny should be streptomycin-dependent. If the gene is chromosomal, one-half should be streptomycin-dependent.

5. Something is transmitted through the high-mother's milk to the female offspring that causes breast cancer. In mice, it is thought to be a virus in the milk. However, the nursing of offspring seems to be required for the appearance of breast cancer, suggesting a hormonal trigger.

CHAPTER

21

Genes and Differentiation

Important Terms and Concepts

Determination is the commitment of a specific cell to a specific developmental fate. **Differentiation** is the phenotypic expression of a cell's determined fate. **Development** represents a programmed sequence of gene activation and inactivation. In the vast majority of species, the DNA content of each cell remains constant.

The partitioning of **cytoplasmic factors** during the early cleavage divisions leads to very early differences between cells, which affect their development.

A **phenocopy** is an environmentally induced phenotype that mimics a genetic mutation. Environmental agents frequently have a **phenocritical period** during development when they are most effective. **Teratogenesis** is a nongenetic process that results in abnormal development.

The maternal genotype can affect the architecture and composition of the egg, thereby affecting the phenotype of the offspring. This process is known as a **maternal effect.**

Totipotent refers to the ability of the nuclei of embryonic cells to support complete development. The degree of totipotency at a specific point in development varies from species to species.

The degree of **regeneration** possible varies from species to species. Regeneration results from differentiated cells maintaining the potential to revert to an embryonic state and undergo differentiation again.

The developing embryo passes through the **blastula** stage into the **gastrula** stage. Totipotency gradually decreases during this process in some species but not in others. Development proceeds by a sequence of gene activation and inactivation.

In *Drosophila*, **puffs** and **Balbiani rings** correspond to RNA transcription. The puffing pattern is tissue specific and developmental-stage specific. Each differentiated tissue has its own set of expressed genes. Some developmental mutants affect the sequence and the timing of gene expression.

Sex determination is achieved in a variety of ways in different species. Sex chromosomes, sex genes, environmental mechanisms, hormonal mechanisms, and a balance between autosomes and X chromosomes are some of these ways.

Cancer is characterized as unregulated growth. It can be caused by changes at the DNA level. Tumor viruses induce malignancy through **oncogenes.** Oncogenes may differ from related host genes by a single base change in the DNA.

Transformation of cells refers to the changes in cell phenotype that occur when the cell becomes malignant. The changes are pleiotropic effects of a small number of DNA lesions.

An **antibody** is a protein that recognizes a specific steriochemical shape determined by an **antigen.** Antibodies are proteins that are classed as **immunoglobulins.** The structure of a specific antibody is unique. The vast number of different antibodies is made possible by **somatic recombination** within **lymphocytes.**

Be sure that you have thoroughly read the entire chapter before you attempt any of the problems.

Solutions to Problems

1. Somatic mutation, mitotic crossing-over, mitotic nondisjunction, mitotic chromosome loss, position-effect variegation as a result of translocation and inversion, cytoplasmic mutation followed by segregation, unrepaired mismatch following recombination, fusion of two zygotes, X chromosome inactivation, transposition.

2. **a.** A phenocopy is an environmentally induced phenotype that closely resembles a phenotype due to a mutation. Proper diagnosis, genetic counseling, and prevention all depend upon distinguishing phenocopies from mutations.

b. By studying the time of action of a teratogenic agent, which produces the phenocopy, the time of action of the gene that produces the similar phenotype can be pinpointed.

c. Phenocopies are not inherited, although they may appear to be inherited if the next generation is exposed to the same teratogenic agent. Mutations are inherited. In animals, genetic crosses will distinguish between the two. In humans, a detailed family medical history may distinguish between the two.

3. Human females are XX; parthenogenesis cannot produce a normal XY male. However, a dominant mutation that causes a sex reversal could occur that would lead to a phenotypically male offspring through parthenogenesis. Such a male would be sterile because he would still be XX. The ovum carrying such a mutation would have to experience nondisjunction at the second meiotic division to produce a diploid zygote.

4. **a.** The basic mechanisms involving gene expression, DNA replication, and cell division are compatible enough to allow for survival.

b. This result indicates that (1) some of the chicken sequences that had been inactivated were not permanently inactivated and possibly could be turned on again in normal chicken erythrocyte cells; (2) inactivation and reactivation have approximately the same mechanism in these two species; and (3) hemoglobin synthesis may be a highly specialized activity that can occur only in the highly specialized erythrocyte cell.

5. Human development, like the development of all multicellular organisms, is a process that continues throughout life and only ends when death occurs. It is a result of the complex interaction of genetics, environment, developmental noise (see Chapter 1), and chance. There is no "final phenotype" per se. A measure of the variability of phenotype due to genetic variation within a population is the broad-sense heritability discussed in Chapter 23. The production of human clones in the

laboratory would be the first step in an experiment that would have to last for the life span of all the clones, and it would be virtually impossible to tease out the various factors impinging upon each of the clones. Therefore, clonal reproduction of humans would likely be useless in determining the relative effects of heredity and environment.

6. a. Regeneration is controlled by the nucleus.

b. The hat-forming substance, produced by the nucleus, is concentrated in the upper part of the stem. An enucleated cell cannot synthesize more of this substance when the upper part of the stem is removed.

c. The hat-forming substance cannot function in the presence of a hat.

7. a. The white eye gene product acts directly in the tissue in which it is made. If this were not the case, then the eye disk from the wild-type larva would develop into a white eye as a result of circulating product from the host.

b. The wild-type host is producing all the enzymes required for eye pigment production. The two mutants are obviously blocked at one or more metabolic steps. The wild-type host is providing the missing components of eye pigmentation, either enzymes or the metabolic products of the defective genes, to the cinnabar and vermilion mutants. These missing components are either diffusing into the transplanted disks or are actively being transported into them, resulting in wild-type eyes.

c. Vermilion mutants are blocked in the transformation of tryptophan to kynurenine. Cinnabar mutants are blocked in the transformation of kynurenine to hydroxykynurenine. Therefore, the cinnabar hosts can provide kynurenine that the vermilion disks need to make normal pigment, but the vermilion hosts cannot provide the hydroxykynurenine that the cinnabar disks need to make normal pigment. Wild-type hosts can provide both substances to the two mutants, so both will develop into wild-type when transplanted into the wild-type host.

8. Determine the average life span of different lines for your organism. Cross the lines and see if life span follows a Mendelian pattern of inheritance.

To check the hypothesis that aging is the result of accumulated somatic mutations, expose a group from each of the lines to low levels of a mutagen and observe the resulting life spans. Compare these values with untreated controls.

The data may be hard to interpret. For example, a mutation that led to improved DNA repair might increase life span. It would seem like a life span gene. However, if aging is due to unrepaired errors and improved DNA repair leads to a longer life span, then this would actually be support for the gradual disruption theory.

9. When an excess of S_1 exists, P_1 is produced, which shuts down operon 2 and allows operon 1 to stay active. When S_1 concentrations fall, the S_2 concentration rises, reversing the situation.

10. Begin with purified DNA and purified histones. Mix the two in the right concentration and check for the presence of nucleosomes by electron microscopy. The nucleosomes do self-assemble. Fractionate the DNA into unique, moderately repetitive, and highly repetitive sequences and mix each fraction with histones. Again, check for nucleosomes. There is no difference in the DNA types in their ability to form nucleosomes.

11. The woman was a heterozygote and X-inactivation randomly inactivated each X chromosome in roughly a 50:50 split. The cells containing the malarial parasite are those in which the normal allele is functioning and the mutant allele is inactivated, and parasite-free cells have a functional mutant allele and a nonfunctional normal allele.

12. The observations suggest that the inactivated X chromosome is either not completely inactivated (it is not) or the information on the inactivated X is needed early in development of the normal female (it is needed). The lack of complete inactivation and the functioning in early development of both X's interferes with normal male development. The Y chromosome has almost no effect upon phenotype.

13. Because the ratio of X's to sets of autosomes does not change, the hexaploid cells will still be female.

14. femaleness

15. P +/*tra* × +/*tra*

F_1 1/4 +/+ (1/8 male, 1/8 female)

1/2 +/*tra* (1/4 male, 1/4 female)

1/4 *tra*/*tra* (1/8 normal male, 1/8 transformed male)

The final sexual ratio is 3/8 female : 5/8 male.

16. The suggestion is that dedifferentiation has occurred in these types of cancers, which in turn suggests that they result from malfunctions in gene regulatory mechanisms.

17. The different tumors have different antigens on their surfaces. Each antigen results from a mutation in a different gene. Prior exposure to a specific set of antigens produces immunity only to that set of antigens.

18. *Culture 1*: two ascospores are homothallic and two are *a*

Culture 2: all ascospores are homothallic

19. The rat GH gene inserted at an unknown location in the mouse chromosomes in 1 percent of the cases. Once inserted, it was capable of being induced by heavy metals to produce larger mice. In the pedigree, the inserted gene is behaving as a dominant gene. The original mouse must have been heterozygous for the insertion because roughly 50 percent of its progeny respond to heavy metals.

Potentially, this mechanism could be used in human gene therapy. Some of the problems are (1) the low efficiency of insertion; (2) random location of insertion; (3) the mutant allele remains in the genome; and (4) insertion could inactivate a gene, causing a new mutation.

20. The molecular basis could be an altered amino-acid sequence in a vital enzyme that, with higher temperature, assumes a thermodynamically more stabile configuration having a different level of enzyme activity. Alternatively, the molecular basis could reside in a vital structural protein. To be dominant, the temperature-sensitive alleles must be interrupting a system that tolerates very little disturbance, such as polymerase function or ribosome structure.

21. Defective muscle proteins would interfere with all of the fly's functioning and would very possibly be lethal. Therefore, select temperature-sensitive mutants. Look for flies that are normal at one temperature but paralyzed at another. *If* the effect is reversible, returning the flies to the permissive temperature would allow breeding experiments and a further definition of the mutation.

22. a. Differences in tissue type are correlated with differences in mRNA.

b. Do a similar competition study using labeled liver mRNA and increasing amounts of unlabeled mRNA from the whole body, minus the liver. The difference between them at maximum competition (minimum amount of label binding) would reflect the liver-specific mRNA.

23. The pool of cytoplasmic RNA is a subset of the nuclear pool, indicating that much RNA never leaves the nucleus. Some of this RNA that never leaves the nucleus is from intron transcripts and both leader and trailer transcripts.

24. Isolate RNA from a very early stage embryo and hybridize it back to each parent separately. If there is a difference in allele activation, the alleles from the female parent will probably be activated earlier than those from the male parent because sperm cell genes are highly condensed while those in the egg are not.

Self-Test

1. You are studying the globin gene in chickens by the use of S1 nuclease, which cleaves only single-stranded DNA. Digestion of pancreatic DNA plus nucleosomes indicates that the globin gene is not susceptible to digestion. Digestion of DNA plus nucleosomes from red blood cells, which are nucleated in the chicken, indicates that several sites in the 5′ control region of the gene are cleaved by S1 nuclease. What do these findings suggest?

2. If the vegetal pole of eggs from the frog *Rana* is irradiated with UV light, the mature frog is normal in every way except that it does not have any germ cells. What does this finding suggest?

3. When mouse cells in culture are exposed to exogenous DNA in the presence of calcium phosphate precipitate, both the DNA and the crystals of calcium phosphate enter the cells by a process that looks like phagocytosis when viewed by electron microscopy. The DNA is intimately associated with the crystals. Within 24 hours, the phagocytic vesicles appear to be undergoing exocytosis, the secretion of the vesicle contents into the surrounding medium. If selection for an introduced marker is applied, a very small portion of the mouse cells will survive, and testing reveals that the selected marker is being expressed. Some cell lines will stably express the selected marker. The key event in this process seems to be the quality of the coprecipitate formed. Why is it that very few cells express the selected marker?

4. In the previous problem, it is stated that some cells stably express the selected marker. That means that expression continues in the absence of selection. When recipient chromosomes are examined in these stable expression lines, frequently one or more chromosomes have new bands that were not there prior to the introduction of the exogenous DNA. The new band locations differ from line to line. What do these observations suggest?

5. When genes are introduced into recipient eukaryotic cells through one of several techniques available, how likely is it that these genes will be regulated properly?

Solutions to Self-Test

1. Cells that do not produce globin have no single-stranded regions, while cells actively producing globin have single-stranded regions. It may be that gene activation is accompanied by single-stranded control regions. Gene activity may also require a looser association of DNA with the nucleosomes.

2. The vegetal pole contains some cytoplasmic factor that is required for germ cell production.

3. Most of the DNA that enters the cells through phagocytosis leaves the cells through exocytosis. The DNA that passes into the nucleus must escape the phagocytic vesicle, perhaps by the calcium phosphate crystals dissolving the vesicle membranes.

4. The suggestion is that stable expression is correlated with the appearance of new chromosome bands. This, in turn, suggests that stable expression is brought about by integration, perhaps through a recombination-like process, of the introduced DNA in the chromosomes. The varying locations of the bands suggest that integration is random.

5. The answer to this question could go one of several ways, and it might vary with the gene being introduced. It is known that position effect does occur in translocated material. A similar effect would most likely occur for introduced genes. Thus, a gene normally functioning within the context of euchromatin would be expected to function at a lower rate when located within heterochromatin, and vice versa. Genes that normally are regulated in cis are unlikely to be regulated in a normal fashion simply because the chance that they are located close to their cis regulator in the new cell is low. Genes that normally are regulated by trans acting factors are more likely to be properly regulated in their new locations.

CHAPTER

22

Genetic Analysis of Development: Case Studies

Important Terms and Concepts

Fate maps show the destinies of the descendants of specific cells during embryogenesis. Cells can also be marked genetically through the use of **mosaics**. In species with an invariant pattern of cell division, it is possible to derive a complete **lineage** of every cell from fertilized egg to the mature adult.

There is a **hierarchy of determinative events** as the embryo goes through development. Cells proceed from **totipotence** through stages that become progressively more restrictive in their developmental potential.

In some species, this progressive restriction of potential is accompanied by a loss of chromosomes. In some species, the loss is accompanied by inactivation of genes.

The **pattern of determination** is apparently established in the cytoplasm of the egg in many species and is under the control of the maternal genotype.

Embryonic genes affecting determination for relatively large segments of the embryo do so within the restrictions imposed by the maternal genotype. Other embryonic genes affecting determination do so within the restrictions imposed for the large segment in which they function.

Be sure that you have thoroughly read the entire chapter before you attempt any of the problems.

Solutions to Problems

1. The studies suggest that muscular dystrophy is caused by defects in the nerves that subsequently cause muscular defects. The parabiosis studies show that the muscular defects cannot be induced in adult muscle. Therefore, the nerve effects on muscles must occur during development.

2. **a.** There must be a diffusable substance produced by the anchor cell that affects development of the six P_n cells. The quantity T_1 is the strongest response, and T_3 represents a lack of response due to a low or absent concentration of the diffusable substance.

b. Remove the anchor cell and the six P_n cells. Arrange the P_n cells in a circle around the anchor cells.

c. All six P_n cells will develop the same phenotype, which will depend on the distance from the anchor cells.

3. **a.** Their phenotype is determined by their location within the developing organism.

b. P1a affects the development of AB derivatives through a diffusable substance.

c. P1b does not affect the development of the AB derivative cells.

d. Almost any ablation or switching experiment would be acceptable here.

4. Compare restriction fragments of homozygous wild-type and Notch. You might also isolate labeled mRNA from homozygous wild-type cells and hybridize it to Notch cells. As a third alternative, use hybridized DNA from controls with DNA from the double deletions that result in homozygous, dead embryos. This might allow you to select DNA on either side of the deletion. This DNA can then be used to chromosome walk across the region.

5. The presence or absence of bristles is affected by only the genotype of the cells that produce bristles and is unaffected by the extent of tissue of a different genotype surrounding it.

6. a. The wild-type flies are attracted to light. Select flies that do not move toward light at the end of a tube.

b. Select for flies that do not fly by placing a poisoned food source high in a fly cage to which they must fly. Alternatively, place mutagenized flies on a window ledge and scare them. The ones that do not fly away are very brave, are very stupid, or are the mutants that you want.

c. Add a tasteless toxic compound to a sugar solution on filter paper in the bottom of the cage. Those flies that cannot taste sugar will not eat it, and those that can taste sugar will eat it and die.

7. The flies may have a structural defect in their legs that prevents normal walking, or they may have a neural defect that results in entanglement of the legs. Nerve and muscle action potentials could be compared in the mutant and wild-type. Electron microscopy could reveal structural abnormalities at that level. Biochemical analysis may reveal either abnormal structural proteins or abnormal enzymatic action.

8. To calculate the distance from a given structure to the locus for hyperkinetic, add the number of mosaics in which the structure is normal and the leg hyperkinetic and the number of mosaics in which the structure is yellow and the leg normal. Divide by the total number of mosaics and multiply by 100.

Leg 1—hyperkinetic:

$$100\%\ (50.5 + 33)/(277 + 33 + 50.5 + 223.5) = 14.3$$

Leg 2—hyperkinetic:

$$100\%\ (69 + 44)/(261.5 + 44 + 69 + 215.5) = 19.1$$

Leg 3—hyperkinetic:

$$100\%\ (82.5 + 46.5)/(250.5 + 46.5 + 82.5 + 215.5) = 21.7$$

Antenna—hyperkinetic:

$$100\%\ (82 + 84.5)/(253 + 84.5 + 82 + 179.5) = 27.8$$

Humeral bristle—hyperkinetic:

$$100\%\ (94 + 66.5)/(241 + 66.5 + 94 + 197.5) = 26.8$$

9. The easiest way to demonstrate maternal inheritance is to do reciprocal crosses.

10. a. In order to determine the temporal pattern of gene action, your experiments must involve isolation of the product from a specific gene and injection of that product at different times into a mutant for that gene. This will allow the determination of the earliest time at which the gene product normally functions to produce wild-type.

b. Spacial patterns can be detected by injecting the wild-type gene product into mutants at different locations.

Self-Test

1. In embryo fusion experiments using mammals, the fused embryos cannot be sexed prior to fusion. When XX and XY embryos, as later determined by chromosome analysis, are fused, all the the embryos develop into males. What does this say about the Y chromosome in mammals? In birds, in which the female is the heterogametic sex, what would be the result of fusion of embryos that are of different sexes?

2. Why are all mammalian females mosaics?

3. When twin cattle share one placenta, all females develop as males if their twin is a male. What does this suggest? How would you test your explanation?

4. If the circulatory systems of two embryonic rats are surgically connected and three kidneys are removed, the growth of the remaining kidney proceeds quite vigorously. What does this suggest? What type of mutants would be useful?

5. Tissues and organs take shape, in part, through differential cell death. If cells destined to die in the normal course of events at stage 24 in the wing of a chick are removed at stage 17 and grafted to a region of muscles, they die on schedule. If they are grafted to the dorsal side of the limb bud, they survive. However, if they are grafted at stage 22 rather than at stage 17 to the dorsal side of the limb bud, they die. What seems to be occurring here? What type of mutants would be useful?

6. Female mammals do not go through estrus or menstruation if their percentage of body fat is too low. Transgenic pigs produced by the injection of the human growth hormone gene into eggs are considerably leaner than controls. This is thought to be because the growth hormone favors the conversion of nutrients to protein rather than fats. The females never go into estrus and are sterile. What is the cause of the sterility, and what approaches might be made to overcome it?

7. Large quantities of specific human gene products can be produced in bacteria. However, some of the gene products normally contain sugar additions that bacteria cannot duplicate. For this reason, researchers are

putting these genes into such animals as cows and sheep, hoping that the gene products will be secreted into the milk of these animals. What is required for production of these gene products in milk?

Solutions to Self-Test

1. The mammalian studies suggest that the Y contains one or more genes that are responsible for sex determination. In birds, if the W chromosome contains one or more genes that determine sex, the fused embryos should develop as females.

2. Because of X-inactivation, which occurs randomly in each cell, each female has two genetically distinct cell lines.

3. The blood of cattle contains some factor that alters development toward the male phenotype. The best guess is that the female tissue is responding to testosterone produced by the male twin. Injection of that hormone into the placenta of developing females should confirm that idea. If so, then the gene coding for testosterone could be cloned and introduced into early embryos by injection. A shift in the sex ratio should be observed. It might be better to do this experiment in mice rather than in cattle, however.

4. Either the remaining kidney is responding to some humoral factor that controls kidney growth or the organ somehow monitors the amount of body to be served and adjusts its size accordingly. Mutants that have altered size for the kidney would be useful.

5. The experiments suggest that, although a "death clock" is set by stage 17 for these cells, it can be turned off by the imposition of external controls in the appropriate environment until stage 22. After stage 22, the clock cannot be turned off.

Mutants that show abnormal development of the wing would be required for any genetic testing.

6. The data suggest that the females are sterile because their body fat content is too low. One approach to overcoming this might be to feed the females a high-fat diet. Another might be to select a strain with very high body fat and use eggs from it for the injections. A third approach would be to attach regulatory sequences to the growth hormone gene that would reduce the level of transcription from the gene or would turn off transcription of the growth hormone gene after a specific developmental stage.

7. The gene would have to be linked to regulatory sequences of a gene that functions only in milk-producing cells. The regulatory sequences that have been tried are from the ß-lactoglobin gene. This has resulted in low secretion of the exogenous gene product into the milk.

CHAPTER

23

Quantitative Genetics

Important Terms and Concepts

The **genotype** can be identified and studied only through its **phenotypic** effects. The study of **genetics** is the study of allelic substitutions that cause **qualitative** differences in the phenotype. However, the actual variation among organisms is usually **quantitative** rather than qualitative. This gives rise to **continuous** variation among members of a species, rather than discrete differences.

Continuous variation is the result of a **norm of reaction** for each genotype and the fact that most traits are controlled by more than one locus. Two individuals with the same genotype can have different phenotypes. Two individuals with different genotypes can have the same phenotype.

Quantitative traits have a **statistical distribution**. This can be presented as a **frequency histogram** or a **distribution function**. A distribution of phenotypes can be described by its **mode**, which is the most frequent class. Some distributions are **bimodal**. The distribution can also be described by the **mean**, the arithmetic average. The **variance** is the spread around the central class. The **standard deviation** is the square root of the variance. Two variables may be described by their **correlation.**

A collection of observations constitutes a **sample** from the **universe** of all observations. This sample may be **biased** or **unbiased.**

The **heritability** of a trait is the proportion of phenotypic variation that can be attributed to genetic variation. The estimates of genetic and environmental variance are specific to the population and environment in which the estimates were made.

Be sure that you have thoroughly read the entire chapter before you attempt any of the problems.

Solutions to Problems

1. Continuous variation can be represented by a bell-shaped curve. Examples are height and weight. Discontinuous variation results in easily classifiable, discrete entities. Examples are red versus white and taster versus nontaster.

2. a. Broad heritability is $H^2 = s_g^2/(s_g^2 + s_e^2)$. Narrow heritability is $h^2 = s_a^2/(s_a^2 + s_d^2 + s_e^2)$, where $s_g^2 = s_a^2 + s_d^2$.

Shank length:

$$H^2 = (46.5 + 15.6)/(46.5 + 15.6 + 248.1) = 0.200$$

$$h^2 = 46.5/(46.5 + 15.6 + 248.1) = 0.150$$

Neck length:

$$H^2 = (73.0 + 365.2)/(73.0 + 365.2 + 292.2) = 0.600$$

$$h^2 = 73.0/(73.0 + 365.2 + 292.2) = 0.010$$

Fat content:

$$H^2 = (42.4 + 10.6)/(42.4 + 10.6 + 53.0) = 0.500$$

$$h^2 = 42.4/(42.4 + 10.6 + 53.0) = 0.400$$

b. The larger the h^2 value, the more that characteristic will respond to selection. Therefore, fat content would respond best to selection.

c. The formula needed is $h^2 \times$ (selection differential) = (selection response). Therefore, selection response = (0.400)(4.0) = 1.6% decrease in fat content, or 8.9% fat content.

3. a. Homozygotes at one locus can be homozygotic at A or at B or at C. The probability of being homozygotic is 1/2 (for A/a: AA or aa), and the probability of being heterozygotic is 1/2. Putting this all together,

$$p(\text{homozygotic at 1 locus}) = 3(1/2)^3 = 3/8$$

$$p(\text{homozygotic at 2 loci}) = 3(1/2)^3 = 3/8$$

$$p(\text{homozygotic at 3 loci}) = (1/2)^3 = 1/8$$

b. p(0 capital letters) = p(all homozygous recessive) = $(1/4)^3 = 1/64$

p(1 capital letter) = p(1 heterozygote and 2 homozygous recessive) = $3(1/2)(1/4)(1/4) = 3/32$

p(2 capital letters) = p(1 homozygous dominant and 2 homozygous recessive) or p(2 heterozygotes and 1 homozygous recessive) = $3(1/4)^3 + 3(1/4)(1/2)^2 = 15/64$

p(3 capital letters) = p(all heterozygous) or p(1 homozygous dominant, 1 heterozygous, and 1 homozygous recessive) = $(1/2)^3 + 6(1/4)(1/2)(1/4) = 10/32$

p(4 capital letters) = p(2 homozygous dominant and 1 homozygous recessive) or p(1 homozygous dominant and 2 heterozygous) = $3(1/4)^3 + 3(1/4)(1/2)^2 = 15/64$

p(5 capital letters) = p(2 homozygous dominant and 1 heterozygote) = $3(1/4)^2(1/2) = 3/32$

p(6 capital letters) = p(all homozygous dominant) = $(1/4)^3 = 1/64$

4. For three genes there are a total of 27 genotypes that will occur in predictable proportions. For example, there are three genotypes that have two heterozygotes and a homozygote recessive (*Aa Bb cc, Aa bb Cc, aa Bb Cc*). The frequency of this combination is $3(1/2)(1/2)(1/4) = 3/16$, and the phenotypic score is 3 + 3 + 1 = 7. The distribution of scores is as follows:

Score	Proportion
3	1/64
5	6/64
6	3/64
7	12/64
8	12/64
9	11/64
10	12/64
11	6/64
12	1/64

5. The population described above would be distributed as follows:

3 bristles	19/64
2 bristles	43/64
1 bristle	1/63

Note that the 3-bristle class contains 7 different genotypes, the 2-bristle class contains 19 different genotypes, and the 1-bristle class contains only 1 genotype. It would be very difficult to determine the underlying genetic situation by doing controlled crosses and determining progeny frequencies.

6. First, solve the formula for values of X over the range for each genotype. Plot those values. Next, add the three values, one for each genotype, for each phenotypic value and plot these summed values. This will give you the overall population distribution.

Phenotypic		Genotype		
value	1	2	3	Total
0.03	0.9037			0.9037
0.04	0.9147			0.9147
0.05	0.9250			0.9250
0.06	0.9347			0.9347
0.07	0.9437			0.9437
0.08	0.9520			0.9520
0.09	0.9597			0.9597
0.10	0.9667		0.9020	1.8687
0.11	0.9730		0.9155	1.8885
0.12	0.9787	0.0000	0.9280	1.9067
0.13	0.9837	0.1900	0.9395	2.1132
0.14	0.9880	0.3600	0.9500	2.2980
0.15	0.9917	0.5100	0.9595	2.4612
0.16	0.9947	0.6400	0.9680	2.6027
0.17	0.9970	0.7500	0.9755	2.7225
0.18	0.9987	0.8400	0.9820	2.8207
0.19	0.9997	0.9199	0.9875	2.8972
0.20	1.0000	0.9600	0.9920	2.9520
0.21	0.9997	0.9900	0.9955	2.9852
0.22	0.9987	1.0000	0.9980	2.9967
0.23	0.9970	0.9900	0.9995	2.9865
0.24	0.9947	0.9600	1.0000	2.9547
0.25	0.9917		0.9995	1.9912
0.26	0.9880		0.9980	1.9860
0.27	0.9837		0.9955	1.9792

0.28	0.9787	0.9920	1.9707
0.29	0.9730	0.9875	1.9605
0.30	0.9667	0.9820	1.9487
0.31	0.9597	0.9755	1.9352
0.32	0.9520	0.9680	1.9200
0.33	0.9437	0.9595	1.9032
0.34	0.9347	0.9500	1.8847
0.35	0.9250	0.9395	1.8645
0.36	0.9147	0.9280	1.8427
0.37	0.9037	0.9155	1.8192
0.38		0.9020	0.9020

The overall population distribution will not result in three distinct modes. With sufficient variation within genotypes, there is a continuous distribution of phenotypes.

7. mean = sum of all measurements/number of measurements

$$= \frac{1 + 2(4) + 3(7) + 4(31) + 5(56) + 6(17) + 7(4)}{1 + 4 + 7 + 31 + 56 + 17 + 4}$$

$$= \frac{154}{120} = 4.7$$

variance = average squared deviation from the mean

$$= \frac{1}{120}\Sigma\begin{bmatrix}(1-4.7)^2 + (2-4.7)^2 + (3-4.7)^2 + (4-4.7)^2 + \\ (5-4.7)^2 + (6-4.7)^2 + (7-4.7)^2\end{bmatrix}$$

$$= \frac{31.43}{120} = 0.2619$$

standard deviation = square root of variance = 0.5117

8.

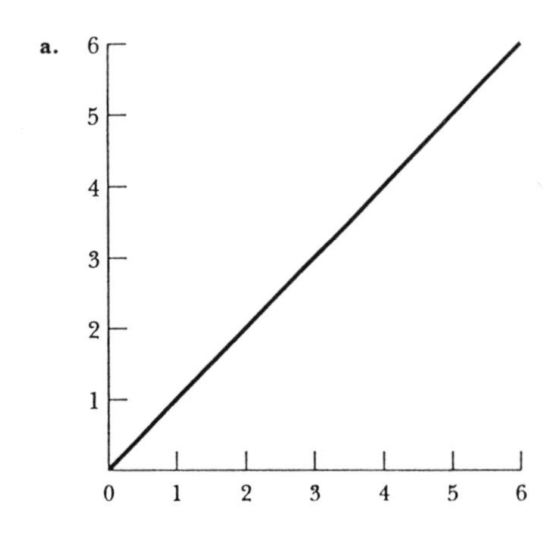

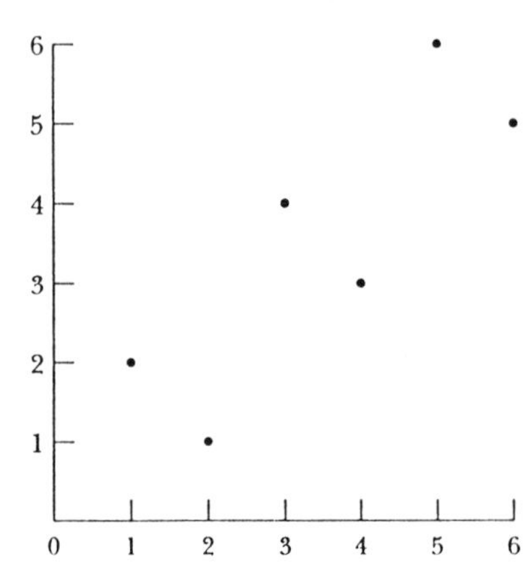

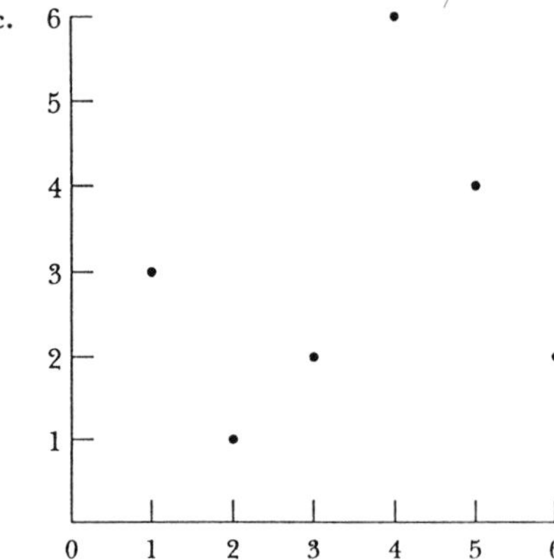

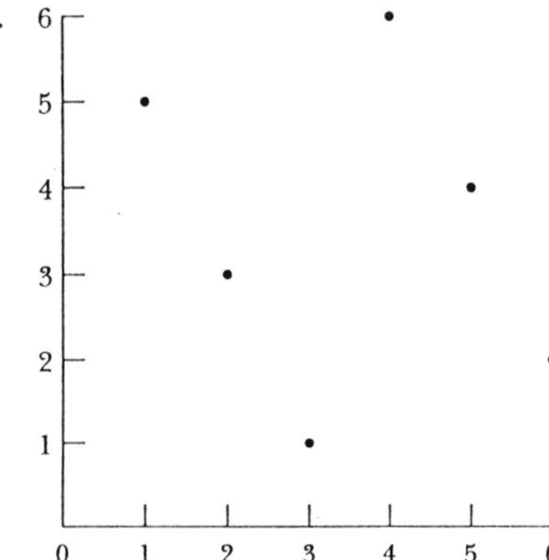

Use the following formula to calculate the correlation between X and Y:

$$\text{correlation} = r_{xy} = \frac{\text{cov } xy}{s_x \, s_y}, \text{ where}$$

$$\text{cov } xy = \frac{1}{N}\Sigma \, x_i y_i - \bar{x} \, \bar{y}$$

a. cov xy = 1/6[(1)(1) + (2)(2) + (3)(3) + (4)(4) + (5)(5) + (6)(6) - (21/6)(21/6)] = 13.125

$$s_x = \sqrt{\frac{1}{N}\Sigma x_i^{\,2} - \bar{x}^2} = \sqrt{\frac{1}{6}\left[1^2 + 2^2 + 3^2 + 4^2 + 5^2 + 6^2 - \left(\frac{21}{6}\right)^2\right]}$$

$= 3.623$

$$s_y = \sqrt{\frac{1}{N}\Sigma x_i^{\,2} - \bar{x}^2} = \sqrt{\frac{1}{6}\left[1^2 + 2^2 + 3^2 + 4^2 + 5^2 + 6^2 - \left(\frac{21}{6}\right)^2\right]}$$

$= 3.623$

Therefore, $r_{xy} = 13.125/(3.623)(3.623) = 1.0$. The other correlation coefficients are calculated in a like manner.

b. 0.83

c. 0.66

d. –0.20

9. a. H^2 has meaning only with respect to the population that was studied in the environment in which it was studied. Otherwise, it has no value.

b. Neither H^2 nor h^2 are reliable measures that can be used to generalize from a particular sample to a "universe" of the human population. They certainly should not be used in social decision making (as implied by the terms "eugenics" and "dysgenics").

c. Again, H^2 and h^2 are not reliable measures and they should not be used in any decision making with regard to social problems.

10. The following are unknown: (1) norms of reaction for the genotypes affecting IQ; (2) the environmental distribution in which the individuals developed; and (3) the genotypic distributions in the populations. Because heritability is specific to a specific population and its environment, even if the above were known, the difference between two different populations cannot be given a value of heritability.

11. First, define alcoholism in behavioral terms. Next, realize that all observations must be limited to the behavior you used in the definition and all conclusions from your observations are applicable only to that behavior. In order to do your data gathering, you must work with a population in which familiarity is distinguished from heritability. In practical terms, this means using individuals who are genetically close but who are found in all environments possible.

12. Before beginning, it is necessary to understand the data. The first entry, *h*/*h h*/*h*, refers to the II and III chromosomes, respectively. Thus, there are four *h* sets of alleles in two or more genes on separate chromosomes. The next entry is *h/l h/h*. Chromosome II is heterozygous and chromosome III is homozygous.

The effect of substituting a low for a high chromosome II can be seen within each row. In the first row, the differences are 25.1 – 22.2 = 2.9 and 22.2 – 19.0 = 3.2. In the second row the differences are 3.1 and 5.2. They are 2.7 and 6.8 in the third row. The average difference is 23.9/6 = 3.98, which actually tells you very little.

The effect of substituting one *l* chromosome for an *h* chromosome in chromosome II, and therefore going from homozygous *hh* to heterozygous *hl*, can be seen in the differences along the rows in the first two columns. The average change is (2.9 + 3.1 + 2.7)/3 = 2.9. When chromosome II goes from heterozygous *hl* to homozygous *ll*, the average change is (3.2 + 5.2 + 6.8)/3 = 5.1.

The effect of substituting one *l* chromosome for an *h* chromosome in chromosome III, and therefore going from homozygous *hh* to heterozygous *hl*, can be seen in the differences between rows (25.1 – 23.0 = 2.1, 22.2 – 19.9 = 2.3, 19.0 – 14.7 = 4.3, 23.0 – 11.8 = 11.2, 19.9 – 9.1 = 10.8, 14.7 – 12.4 = 12.4). When chromosome III goes from homozygous *hh* to heterozygous *hl*, the average change is (2.1 + 2.3 + 4.3)/3 = 2.9. When it goes from heterozygous *hl* to homozygous *ll*, the average change is (11.2 + 10.8 + 12.4)/3 = 11.5.

A summary of these results appears below:

	Chromosome II	Chromosome III	Total
hh to *hl*	2.9	2.9	5.8
hl to *ll*	5.1	11.5	16.6

Now it should be clear that each set of alleles for both chromosomes is expressed in the phenotype, but that expression varies with the chromosome. Chromosome III appears to have a stronger affect on the phenotype than chromosome II (compare total amount of change). There is some dominance of *h* over *l* for both chromosomes because the change from *hh* to *hl* is less than the change from *hl* to *ll*. Finally, there is definitely some epistasis occurring. Compare *h/h h/h* with both *l/l h/h* and *h/h l/l*. The difference in the first case is 6.0 and, in the second case, 13.3. The expected amount of change in going from *h/h h/h* to *l/l l/l* is therefore 6.0 + 13.3 = 19.3. The *l/l l/l* phenotype should be 25.1 – 19.3 = 5.8, but the observed value is 2.3.

13.

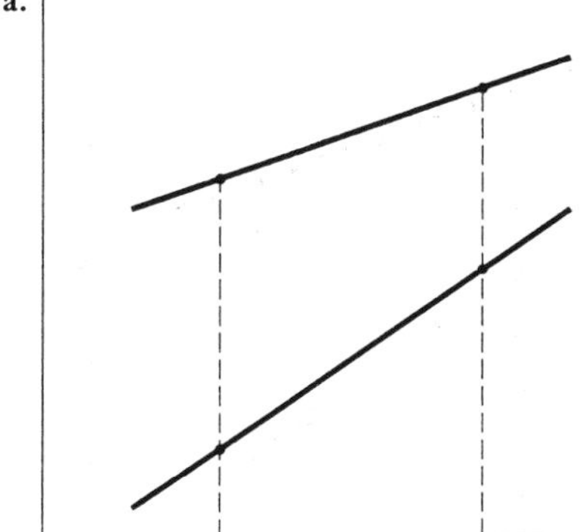

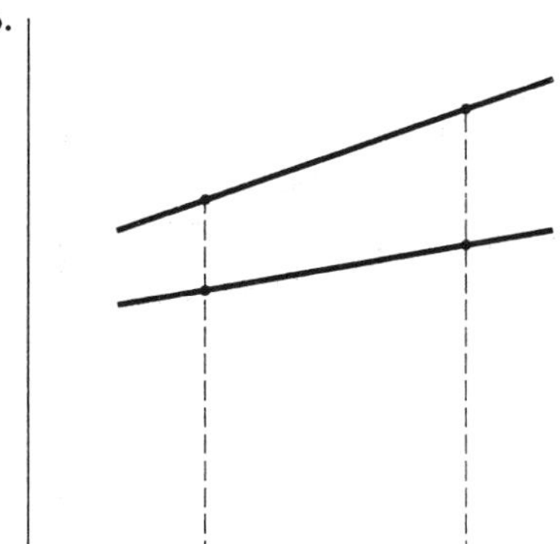

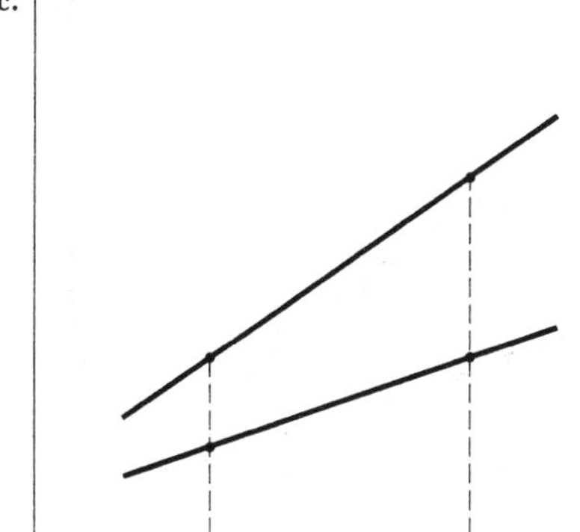

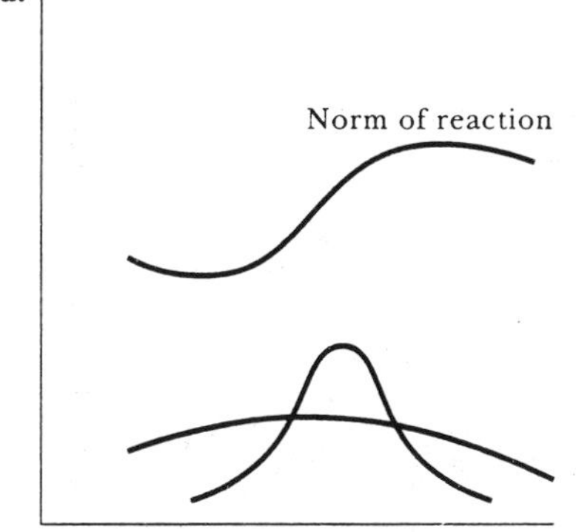

14. Assume that $s_p^2 = s_x s_y$. Then

$$H^2 = 4(\text{correlation of full sibs}) - 2(\text{parent} - \text{offspring correlation})$$

$$= \frac{4\,(\text{cov full sibs})}{s_p^2} - \frac{2\,(\text{cov parent} - \text{offspring})}{s_p^2}$$

$$= \frac{(4)\,(2.7)}{14.8} - \frac{(2)(1.7)}{14.8} = 0.5002$$

However, H^2 can also be calculated as the difference between full and half-sibs. The formula for that is

$$\frac{4\,(\text{cov full sibs} - \text{cov half-sibs})}{s_p^{\,2}}$$

When the proper numbers are put into the formula, $H^2 = 0.5135$. The average value of H^2 is $(0.5135 + 0.5002)/2 = 0.5068$.

To calculate heritability in the narrow sense, use the following formula:

$$h^2 = \frac{2\,(\text{cov parent} - \text{offspring})}{s_p^{\,2}}$$

$$= \frac{(2)(1.7)}{14.8} = 0.2297$$

15. a. If you assume that individuals at the extreme of any spectrum are homozygous, then their offspring are more likely to be heterozygous than the original individuals. That is, they will be less extreme.

b. For Dalton's data, regression is an estimate of heritability (h^2), assuming that there were few environmental differences between father and son.

Self-Test

1. In a study of body weight among Labrador retrievers, the total variance is 4.80. The covariance between half-siblings is 0.76. Estimate the narrow-sense heritability of body weight in this population.

2. Two inbred lines of sheep are intercrossed. The variance in the weight of the F_1 is 3.2. The F_1 are crossed, and the variance in the F_2 is 7.3. What is the estimate of broad-sense heritability in sheep weight?

3. In Maine, there is a fairly high frequency of cats with abnormal numbers of paws. Calculate the mean, variance, and standard deviation of the distribution of paws observed in one community. Does it make sense to characterize the observations in this fashion?

Paw number	Number of individuals
4	93
5	0
6	4
7	0
8	3

4. The accident rate among human males aged twenty is much higher than among females. List some of the hypotheses that might be tested experimentally.

5. The number of eye facets in *Drosophila* was measured at various temperatures. What is the correlation between numbers of facets and temperature?

Temperature, $^{\circ}$C	Facets
15	200
20	150
25	90
30	70

Solutions to Self-Test

1. The covariance for half-siblings is 1/4 of the additive genetic variance (Table 23-2):

$$s_a^2 = 4 \times \text{cov} = (4)(0.76) = 3.04$$

Heritability in the narrow sense is

$$h^2 = \frac{s_a^2}{s_p^2}$$

so that $h^2 = 3.04/4.80 = 0.63$, or 63%.

2.

$$s_e^2 = 3.2$$

$$s_e^2 + s_g^2 = 7.3$$

$$s^2 = 7.3 - 3.2 = 4.1$$

$$H^2 = 4.1/7.3 = 0.56, \text{ or } 56\%$$

3. mean = (4)(93) + (6)(4) + (3)(8) = 4.2

variance = $s^2 = 1/100[(4-4.2)^2 + (5-4.2)^2 + (6-4.2)^2 + (7-4.2)^2 + (8-4.2)^2 = 0.262$

standard deviation = s = 0.512

Although the population can be categorized in this fashion, the number of paws on a cat is not a continuous distribution. It would make more sense to study this trait by traditional Mendelian methods.

4. Many hypotheses could be proposed. The following are a subset of them.

Males are genetically more susceptible to accidents than females.

Males are environmentally conditioned to take more risks than females.

There is a familial component to risk-taking among males and females. This component varies among families.

Males have poorer reaction times than females.

Males have poorer vision than females.

5. mean temperature, $\bar{x} = 22.5$

mean facets, $\bar{y} = 121.7$

standard deviation temperature, $s_x = 20.27$

standard deviation facets, $s_y = 121.7$

$\text{cov}_{xy} = -281.25$

correlation $= r_{xy} = -0.114$

CHAPTER

24

Population Genetics

Important Terms and Concepts

Among individuals in a population, there is **phenotypic variation**, or **polymorphism**. Offspring are phenotypically closer to their parents than they are to unrelated individuals. Some phenotypes **survive** to reproduce better than other phenotypes in a given environment. **Natural selection** of these more successful phenotypes in a given environment leads to a **reproductive advantage**, or an increased **fitness**, for them. This results in a change of allelic frequency. **Evolution** is a change in genotypic frequencies.

All variation ultimately comes from **mutation**.

The frequency of a given allele in a specific population is affected by recurrent mutation, selection, migration, and random sampling effects. In an idealized, randomly interbreeding population not subjected to any forces that alter genotypic frequencies, the genotypic frequencies do not change. They can be represented by the **Hardy-Weinberg equilibrium** equation, $p^2 + 2pq + q^2 = 1.0$.

Be sure that you have thoroughly read the entire chapter before you attempt any of the problems.

Solutions to Problems

1. The frequency of an allele in a population can be altered by selection, mutation, migration, inbreeding, and random genetic drift.

2. There are a total of (2)(384) + (2)(210) + (2)(260) = 1708 alleles in the population. Of those, (2)(384) + 210 = 978 are *A1* and 210 + (2)(260) = 730 are *A2*. The frequency of *A1* is 978/1708 = 0.57, and the frequency of *A2* is 730/1708 = 0.43.

3. The given data are $q^2 = 0.04$ and $p^2 + 2pq = 0.96$. If $q^2 = 0.04$, $q = 0.2$ and $p = 0.8$. To check this, use these numbers in the second equation: $(0.8)^2 + (2)(0.8)(0.2) = 0.64 + 0.32 = 0.96$. The frequency of *BB* is 0.64, and the frequency of *Bb* is 0.32.

4. This problem assumes that there is no backward mutation. Use the following equation: $p_n = p_o e^{-n\mu}$. That is,

$$p_{50,000} = (0.8)e^{-(5\times10^4)(4\times10^{-6})} = (0.8)(0.81873) = 0.65$$

5. a. If the variants represent different alleles of gene *X*, a cross between any two variants should result in a 1:1 progeny ratio. All the variants should map to the same locus. Amino acid sequencing of the variants should reveal differences of one to just a few amino acids.

b. There could be another gene (gene *Y*), with five variants, that modifies the gene *X* product posttranscriptionally. If so, the easiest way to distinguish between the two explanations would be to find another mutation in *X* and do a dihybrid cross. For example, if there is independent assortment,

P	$X^1\ Y^1 \times X^2\ Y^2$
F_1	$1\ X^1\ Y^1 : 1\ X^1\ Y^2 : 1\ X^2\ Y^1 : 1\ X^2\ Y^2$

If the mutation in *X* led to no enzyme activity, the ratio would be

2 no activity : 1 variant 1 activity : 1 variant 2 activity

The same mutant in a one-gene situation would yield 1 active : 1 inactive.

6. a. If the population is in equilibrium, $p^2 + 2pq + q^2 = 1$. Use p from the data to predict the frequency of q, and then check the calculated values against the observed.

$$p = [406 + 1/2(744)]/1482 = 0.5249$$

$$q = 1 - p = 0.4751 = \text{predicted value}$$

The phenotypes should be distributed as follows if the population is in equilibrium.

$$L^M L^M = p^2(1482) = 408$$

$$L^M L^N = 2pq(1482) = 739$$

$$L^N L^N = q^2(1482) = 334$$

b. If mating is random with respect to blood type, then the following frequency of matings should occur.

$$L^M L^M \times L^M L^M = (p^2)(p^2)(741) = 56.25$$

$$L^M L^M \times L^M L^N = (2p^2)(2pq)(741) = 203.6$$

$$L^M L^M \times L^N L^N = (2p^2)(q^2)(741) = 92$$

$$L^M L^N \times L^M L^N = (2pq)(2pq)(741) = 184.28$$

$$L^M L^N \times L^N L^N = (2)(2pq)(q^2)(741) = 166.8$$

$$L^N L^N \times L^N L^N = (q^2)(q^2)(741) = 37.75$$

The mating is random with respect to blood type.

7. a. When the allelic frequency differs between sexes for an X-linked gene, $p = 1/2(1/3p_m + 2/3p_f)$. At generation 0, $p = 1/2[1/3(0.8) + 2/3(0.2)] = 0.2$.

In generation 1, all males get their X from their mother, and therefore, the frequency of p in the males will be the same as it is in the mother, 0.2. All females get an X from each parent at the allelic frequency in each parent. Therefore, the daughters will have a frequency of $p_1 = (1/2)(p_m + p_f) = (1/2)(0.8 + 0.20) = 0.5$. The following table provides information for further generations.

Generation	p male	p female
0	0.8	0.2
1	0.2	0.5
2	0.5	0.35
3	0.35	0.425
.	.	.
.	.	.
n	$p_{f(n-1)}$	$p_{1/2[m(n-1)+f(n-1)]}$

where m = male and f = female.

b. Let p = frequency in males and p' = frequency in females. For any generation, $p_n = p'_{n-1}$ and $p'_n = (1/2)(p_{n-1} + p'_{n-1})$. The difference between these two, d, is

$$d = (1/2)(p_{n-1} + p'_{n-1}) - p'_{n-1}$$
$$= (1/2)(p_{n-1} - p'_{n-1})$$

Given the initial values of p_0 and p'_0,

$$d_n = (1/2)^n(p_0 - p'_0)$$

8. **a.** and **b.**

Population	p	q	Equilibrium?
1	1.0	0.0	yes
2	0.5	0.5	no
3	0.0	1.0	yes
4	0.625	0.375	no
5	0.3775	0.62	no
6	0.5	0.5	yes
7	0.5	0.5	no
8	0.2	0.8	yes
9	0.8	0.2	yes
10	0.993	0.007	yes

c. $4.9 \times 10^{-6} = 5 \times 10^{-6}/s$; $s = 0.102$

d.

Genotype	Frequency	Fitness	Gametes	A	a
AA	0.25	1.0	0.25	0.25	0.0
Aa	0.50	0.8	0.40	0.20	0.20
aa	0.25	0.6	0.15	0.0	0.15
				0.45	0.35

$$p = 0.45/(0.45 + 0.35) = 0.56$$

$$q = 0.35/(0.45 + 0.35) = 0.44$$

9. **a.** Assuming equilibrium, if $q = 0.1$, $q^2 = 0.01$.

b. 10 times

c. Marriages in which one-half of the children of both sexes would be color blind are X^BX^b X X^bY. Such marriages occur with a frequency of $(2pq)(q) = 2pq^2 = 2(0.9)(0.1)^2 = 0.018$.

d. All children would be normal if the female were homozygous normal. The frequency of such marriages is $p^2(p + q) = (0.9)^2(0.5 + 0.5) = 0.81$.

e. Color blind females result from two types of matings:

$$X^BX^b \text{ X } X^bY = (2pq)(q') = 2(0.2)(0.8)(0.6) = 0.192$$

$$X^bX^b \text{ X } X^bY = (q)^2(q') = (0.2)^2(0.6) = 0.024$$

Half the females from the first mating and all the females from the second will be color blind, so the frequency of color blind female progeny is $= (1/2)(0.192) + 0.024 = 0.12$.

Color blind males will result when the mother is either heterozygous (half the male offspring) or homozygous recessive (all the male offspring), regardless of the father's genotype. Therefore, the frequency of color blind male progeny is

$$(1/2)(2pq) + q^2 = (1/2)(2)(0.8)(0.2) + (0.2)^2 = 0.2$$

f. The male frequency will be 0.2 (frequency of the female in the previous generation) and the female frequency will be $(1/2)(0.2 + 0.6) = 0.4$.

10. The frequency of a phenotype in a population is a function of the frequency of alleles that lead to that phenotype in the population. To determine dominance and recessiveness, do standard Mendelian crosses.

11. Assume that proper function results from the right gene products in the proper ratio to all other gene products. A mutation will change the gene product, eliminate the gene product, or change the ratio of it to all other gene products. All three outcomes upset a previously balanced system. While a new balance may be achieved, this is unlikely.

12. Dominance is usually wild-type because most detectable mutations in enzymes result in lowered or eliminated enzyme function. To be dominant, the heterozygote has approximately the same phenotype as the homozygote dominant. This will be true only when the wild-type allele produces a product and the mutant allele does not.

The chromosomal rearrangements are dominant mutations because so many genes are affected that it is highly unlikely that all of their alleles will be dominant and "cover" for them.

13. Prior to migration $q^A = 0.1$ and $q^B = 0.3$. Immediately after migration, $q^{A+B} = (1/2)(q^A + q^B) = (1/2)(0.1 + 0.3) = 0.2$. The frequency of affected males is 0.2 and the frequency of affected females is $(0.2)^2 = 0.04$.

14. The probability of homozygous by descent (f) is $f = (1/2)^n$, where n = number of ancestors in a closed loop.

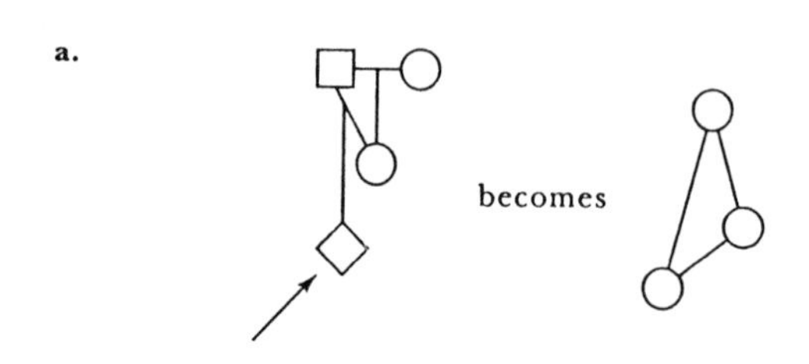

a. $f = (1/2)^3 = 1/8$

b.

becomes

b. $f = (1/2)^5 + (1/2)^5 = 1/16$

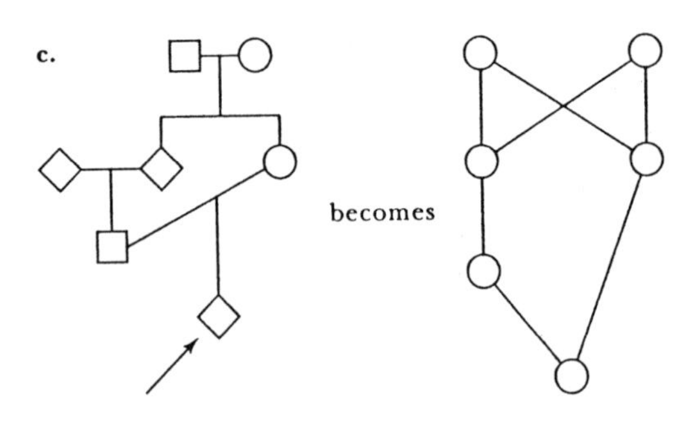

c. $f = (1/2)^4 + (1/2)^4 = 1/8$

15. Albinos may have been considered lucky and encouraged to breed at very high levels in comparison to nonalbinos. They may also have been encouraged to mate with each other. Alternatively, in the tribes with a very low frequency, albinos may have been considered very unlucky and destroyed at birth or prevented from marriage.

16. The allele frequencies are

$$A\text{: } 0.2 + (1/2)(0.60) = 50\%$$

$$a\text{: } (1/2)(0.60) + 0.2 = 50\%$$

Positive assortive mating: The alleles will randomly unite within a phenotype. For A—, the mating population is 0.2 AA + 0.6 Aa. The allele frequencies within this population are

A: $[0.2 + (1/2)(0.6)]/0.8 = 0.625$

a: $(1/2)(0.6)/0.8 = 0.375$

The phenotypic frequencies that result are

A–: $p^2 + 2pq = (0.625)^2 + 2(0.625)(0.375) = 0.3906 + 0.4688 = 0.8594$

aa: $q^2 = (0.375)^2 = 0.1406$

However, because this subpopulation represents 0.8 of the total population, these figures must be adjusted to reflect that by multiplying by 0.8:

$$A\text{–: } (0.8)(0.8594) = 0.6875$$

$$aa\text{: } (0.8)(0.1406) = 0.1125$$

The aa contribution from the other subpopulation will remain unchanged because there is only one genotype, aa. The contribution to the total phenotypic frequency is 0.20. Therefore, the final phenotypic frequencies are A– = 0.6875 and aa = 0.20 + 0.1125 = 0.3125. These frequencies will remain unchanged over time, but the end result will be two separate populations, AA and aa, which will not interbreed.

Negative assortive mating: If assortive mating is between unlike phenotypes, the two types of progeny will be Aa and aa. AA will not exist. Aa will result from all $AA \times aa$ matings and one-half of the $Aa \times aa$ matings. The matings will occur with the following frequencies:

$AA \times aa = (0.2)(0.2) = 0.04$

$$Aa \times aa = (0.6)(0.2) = 0.12$$

Because these are the only matings that will occur, they must be put on a 100 percent basis by dividing by the total frequency of matings that occur:

$AA \times aa$: 0.04/0.16 = 0.25, all of which will be Aa

$Aa \times aa$: 0.12/0.16 = 0.75, half Aa and half aa

The phenotypic frequencies in the next generation will be

Aa: 0.25 + 0.75/2 = 0.625

aa: 0.75/2 = 0.375

In the second generation, the same method will result in a final ratio of 0.5 Aa : 0.5 aa. These values will remain unchanged after the second generation of negative assortive mating.

17. Many genes affect bristle number in *Drosophila*. The artificial selection resulted in lines with mostly high-bristle-number alleles. Some mutations may have occurred in 20 generations, but most of the response was due to alleles present in the original population. Assortment and recombination generated lines with more high alleles.

Fixation of some alleles causing high bristle number would prevent complete reversal. Some high-bristle-number alleles would have no bad effects on fitness, so there would be no force pushing bristle number back down due to those loci.

The low fertility in the high line could have been due to pleiotropy or linkage. Some alleles that caused high bristle number may also have caused low fertility (pleiotropy). Chromosomes with high-bristle-number alleles may also carry alleles at different loci that caused low fertility (linkage). After artificial selection was relaxed, the low-fertility alleles would have been selected against through natural selection. A few generations of relaxed selection would have allowed low-fertility-linked alleles to recombine away, producing high- bristle-number chromosomes that did not contain low-fertility alleles. When selection was reapplied, the low-fertility alleles were reduced in frequency or separated from the high-bristle loci, so this time there was much less of a fertility problem.

18. a. The needed equation is $p' = p\,\overline{W}_a / \overline{W}$, where $\overline{W}_a = pW_{AA} + qW_{Aa}$ and $\overline{W} = p^2 W_{AA} + 2pqW_{Aa} + q^2 W_{aa}$.

$$p' = \frac{(0.5)[(0.5)(0.9) + (0.5)(1.0)]}{(0.5)^2(0.9) + 2(0.5)(0.5)(1.0) + (0.5)^2(0.7)} = 0.528$$

b. $\hat{p} = (W_{aa} - W_{Aa})/[(W_{aa} - W_{Aa}) + (W_{AA} - W_{Aa})]$

$$= \frac{0.7 - 1.0}{(0.7 - 1.0) + (0.9 - 1.0)} = 0.75$$

19. The equation needed is $q = \sqrt{\mu/s}$ or $s = \mu/q^2 = 10^{-5}/0.001 = 0.01$.

20. Affected individuals = $Bb = 2pq = 4 \times 10^{-6}$. Because q is almost equal to 1.0, $2p = 4 \times 10^{-6}$. Therefore, $p = 2 \times 10^{-6}$.

$$\mu = hsp = (1.0)(0.7)(2 \times 10^{-6}) = 1.4 \times 10^{-6}$$

21. The probability of not getting a recessive lethal genotype for one gene is 1 - 1/8 = 7/8. If there are n lethal genes, the probability of not being homozygous for any of them is $(7/8)^n = 13/31$. From log tables, $n =$ 6.5, or an average of 6.5 recessive lethals in the human genome.

22. a. $\hat{q} = \sqrt{\mu/s} = \sqrt{\frac{10^{-5}}{0.5}} = 4.47 \times 10^{-3}$

$$sq^2 = 0.5(4.47 \times 10^{-3})^2 = 10^{-5}$$

b. $\hat{q} = \sqrt{\mu/s} = \sqrt{\frac{2 \times 10^{-5}}{0.5}} = 6.32 \times 10^{-3}$

$$sq^2 = 0.5(6.32 \times 10^{-3})^2 = 2 \times 10^{-5}$$

c. $\hat{q} = \sqrt{\mu/s} = \sqrt{\frac{10^{-5}}{0.3}} = 5.77 \times 10^{-3}$

$$sq^2 = 0.3(5.77 \times 10^{-3})^{-2} = 10^{-5}$$

Self-Test

1. In a population, the alleles A and a are at initial frequencies of p and q. Prove that the gene frequencies and the zygotic frequencies do not change from generation to generation as long as no forces for change are acting on the population.

2. At what allelic frequency is the heterozygous genotype frequency one-half the homozygous recessive genotype frequency?

3. A strongly odorous substance, mathanethiol, is secreted in some humans. Secretion is recessive. If the frequency of the secretor allele, *m*, is 0.6, what is the probability of a mating between two heterozygous nonsecretors?

4. In the previous problem, what is the probability among all matings of having a secreting girl?

5. In humans, an index finger shorter than the ring finger is autosomal dominant in males and recessive in females. Of 1000 males in a population, 510 had a short index finger. What is the expected frequency of long and short index fingers in 1000 females from this population?

6. If the fitness of a population is 0.85 and the mutation rate to the recessive allele is 6×10^{-6}, what is the equilibrium frequency?

Solutions to Self-Test

1. Random mating results in the following zygotic frequencies:

$$p^2(AA) + 2pq(Aa) + q^2(aa) = 1.0$$

All the gametes of *AA* individuals and half the gametes of *Aa* individuals will be *A*. The frequency of *A* in the next generation will be

$$p^2 + pq = p^2 + p(1 - q) = p^2 + p - p^2 = p$$

Therefore, the frequency of all *a* alleles in the next generation will be $1 - p = q$.

2.
$$\begin{aligned} 1/2(q^2) &= 2pq \\ q^2 &= 4pq \\ &= 4q(1-q) \\ &= 4q - 4q^2 \\ 0 &= 4q - 5q^2 \end{aligned}$$

Either $q = 0$, which is not correct, or $0 = 4 - 5q$, $q = 4/5 = 0.8$, and $p = 0.2$.

3. The frequencies are $q = 0.6$ and $p = 0.4$. The heterozygous population is $2pq$. The frequency of matings between heterozygotes is $(2pq)(2pq) = 4p^2q^2 = 4(0.16)(0.36) = 0.23$.

4. A secretor could result from the following matings:

$$Aa \times Aa \rightarrow 1/4\ aa$$
$$Aa \times aa \rightarrow 1/2\ aa$$
$$aa \times Aa \rightarrow 1/2\ aa$$
$$aa \times aa \rightarrow \text{all } aa$$

The probability of a girl is 1/2. The probability of a secreting girl is $(1/2)(1/4)(2pq)(2pq) + 2(1/2)(1/2)(2pq)(q^2) + 1/2(q^2)(q^2)$

$$= (1/2)(1/4)(2)(0.4)(0.6)(0.4)(0.6) = 0.029$$
$$2(1/2)(1/2)(2)(0.4)(0.6)(0.36) = 0.086$$
$$(1/2)(0.36)(0.36) = \underline{0.065}$$
$$0.180$$

5. The frequencies are $q^2 = 490/1000$, $q = 0.7$, and $p = 0.3$. Among females, however, p is recessive. Therefore, 1000 p^2 females will have a ring finger that is shorter than the index finger, or 90. The remaining 910 females will have a long index finger.

6. $\hat{q} = \sqrt{\mu/s}$, $q = 6 \times 10^{-6}$, $s = 1 - W = 1.0 - 0.85 = 0.15$

$$\hat{q} = \sqrt{\frac{6 \times 10^{-6}}{0.15}} = 0.0063$$

$$p = 0.9937$$